Was There Marijuana in the Garden of Eden?

An Exposé on Marijuana and Hemp

Bruce Meland

The information about marijuana and hemp provided in this book by authors and references contains health related information, but is not meant or should be used to diagnose or treat any medical condition. For diagnosis and treatment of a medical condition consult a qualified physician or health care professional, knowledgeable on such medical matters. Information contained herein is for informational purposes only and do not constitute any endorsement. Readers should also be aware that references and websites in this book may have changed or be incomplete or may have been deleted.

ISBN: Book 978-0-692-16387-0

Cover design by: Robin Locher

Editing by: Barbara Overen Smith

To order additional copies of this book, contact:

Bruce Meland
etimes@teleport.com

Photo Contents

Charlotte with high CBD-rich Charlotte's Web™ Cannabis. Courtesy of www.theroc.us

https://images.search.yahoo.com/images/view

Hoodie

made with hemp Eco Lego Building Blocks

Contents

FORWARD

This publication is an attempt to inform the readers about the history and up to date information about marijuana and hemp. I have divided this book into 2 sections: Marijuana and Hemp.

Evidence of hemp was found in China 10,000 years ago. There were marijuana seeds found in a Asian Shamans Tomb 2700 years ago. It is obvious at least in my mind that animal and human kind were created with hemp and marijuana, or evolved with hemp and marijuana. Also consider the fact that scientists have discovered about 90 different hemp cannabinoid molecules or Phyto cannabinoids (plant cannabinoids) in marijuana and hemp plants and also found about 90 endo cannabinoids or cannabinoid receptors in the human body, termed endocannabinoids. This indicates a organic symbiotic relationship between the plant kingdom and animal kingdom from their beginnings.

Also Marijuana Phyto cannabinoids have dominant THC, tetrahydrocannabinol molecules that generally create a euphoria or high; and in hemp CBD, cannabidiol molecules that have healing or preventive effects on epilepsy, cancer, sleep disorders, PTSD, Parkinson's, diabetes, MS, pain reducer and other nervous disorders. Treating diseases with natural and organic THC and CBD molecules and terpenes have little side effects. As marijuana and hemp research evolves, patients will be able to go online to get updated information as to best cannabis plant to grow, then process into oil and treat themselves.

Rick Simpson's formula for treating cancer is found on his website; see page 19. Patients can be treated with cannabis with the aid of qualified healthcare professionals. Wide use of a variety of cannabinoids, CBD, oils, crystals, and terpenes can achieve these goals. CBD methods of treatment can reduce healthcare costs world wide.

Check out Cannabinoid & Terpene Profile, page 48, and Therapeutic Effects of Cannabinoids, page 49, for more details.

Bruce Meland

BRIEF HISTORY OF MARIJUANA

Charlotte's Web™ Courtesy of www.theroc.us

There was Marijuana in the Garden of Eden. Marijuana thrived in many ancient civilizations. Recorded history shows that Marijuana and Hemp--collectively known as Cannabis, was present in Taiwan almost 12,000 years ago. The plant was referred as Ma in Mandarin Chinese history. Originally, Marijuana and Hemp were one and the same plant but, plant breeding throughout the centuries has produced plants that have become two separate strains of cannabis. Around 100 BC the habit of smoking 'pot' was evident in South Central Asia.

http://mercatia.wikidot.com/hempweaving

(The ancient bronze script for ma "cannabis" is depicted above the ma plants hanging in a shed. http://en.wikipedia.org/wiki/Hemp)

From Taiwan and China, cannabis spread to India, Afghanistan, Pakistan, and the Middle East. Did you notice marijuana was growing around Bin Ladin's compound in Pakistan which was exposed to the world after US Special Forces stormed his compound? From the Mid East cannabis apparently spread to North Africa, then to Europe after Napoleon invaded Egypt. Apparently Napoleon and his army brought cannabis back to France and Europe. Finally cannabis spread to North and South America with the early explorers and settlers.

Magu, Goddess of Longevity, 18th century hanging scroll. Source: National Museum, Warsaw

http:// www.mnw.art.pl/

MaGu literally means 'Hemp Maid.' This legendary painting of the Taoist's immortal Goddess was probably painted on a scroll of silk. MaGu is associated with the elixir of life in Chinese literature. MaGu, as a beautiful young woman, is also a symbol and protector of females in Chinese mythology.

Around 2900 BC, Chinese Emperor Fu, who is credited with bringing civilization to China, made reference to Ma. Around 2700 BC. Chinese Emperor Chen Nong, considered to be The Father of Chinese Medicine, noted Ma's healing properties as well as its Yin and Yang effects. Ma was used along with Chinese herbal medicines ginseng and ephedra to treat ailments. The earliest recorded reference to Ma was found in the 15th century BC Chinese Pharmacopeia, the Rh-Ya. The first edition was published in 1930.

2015 Chinese Pharmacopoeia by China Food & Drug Admin. http://eng.sfda.gov.cn/WS03/CL0757/122060.html

Several Pharaohs including Ramses II, Egypt's most powerful king, 1213 B.C. and Tutankhomen, King Tut, Egypt's most popular King, brought cannabis pollen with them to their grave, found in their wrappings. © Junichi Shimazakil | www.dreamstime.com

Around 1450 BC, holy anointing oil wasdescribed in the original Hebrew version of Exodus (30; V23-24) where cannabis wasextracted into 6 quarts of olive oil along with other fragrant herbs. This potent ointment was used for anointing purposes. In Egypt, cannabis pollen was found on the mummy of Ramses II who died in 1213 BC. Besides creating euphoria with its psychoactive agent, Cannabis was used for treating ailments such as rheumatism, gout, malaria, labor pains, nausea, glaucoma and wounds. (Marihuanawww.druglibrary.org.)

Genesis 1: V.29 And God said, Behold, I havegiven you every herb bearing seed which is on the face of all the earth, and every tree, which is the fruit of a tree yielding seed; to you it shall be for meat. Old Testament, King James Version.

Psalms 104 V.14 He causeth the grass to grow for the cattle and herb for service of man. Abel 19:80 in the Jewish Talmud mentions the euphoric properties of cannabis.

In 7th Century B.C. Persian Prophet, Zoroaster, wrote the "Zend-Avesta," similar to the Indus "Vedas." In both religions, Bhang, a powerful cannabis by-product, was used as a metaphysical narcotic and medicine.

Around 600 BC cannabis was used to treat various health problems in India. They believed it could quicken the mind, prolong life, lower fevers, induce sleep, improve judgment and cure dysentery. It was also used for recreational benefits. The treatise of Indian medicine, Sushruta Samhita, contained various cures using cannabis. (Jon Green, Cannabis, 2002)

Siddhartha Gautama (the Buddha), founder of Buddhism, was born a Kshatriva Warrior Prince in Lumbini Nepal around 400 B.C. His teachings spread to the Ganges Valley and to North East India. In times of famine Buddha, his followers and residents of the area survived on hemp seeds. Legend says that the Buddha lived on a ration of one hemp seed per day during his 6 years of asceticism. Hemp was used for food, fiber and medicine in that area from 400 BC to present times.

DIOSCORIDIS
ANAZARBEI, DE

LVGDVNI.
Apud Balthazarem Arnolletum.

Cover of De Materia Medica wikipedia.org/wiki/Dioscorides

From 40-70 AD Pedanius Dioscorides, a Greek physician, botanist and Roman army doctor, traveled extensively throughout the Roman Empire and published a Book "On Medical Matters" which became the medical guide for the next 1500 years. Importantly, it included information on Cannabis (Martin Booth, Cannabis: A History. 2005.)

Hua Tuo was a Chinese physician who lived in the late Eastern Han Dynasty, 140-208 AD, and was the first Physician to use anesthesia during surgery. He used wine with an herbal concoction called "cannabis boil powder." Besides being respected in surgery and anesthesiology, Hua Tuo was famous for his skills in acupuncture, moxibustion, herbalmedicine, and medical Daoyin exercises. He also developed Wuqinxi, the developed Wuqinxi, the "Exercise of the five animals," from studying the movements of the tiger, deer, bear, ape and crane.

Hua Tuo Chinese surgeon. wikipedia.org/wiki/Hua_Tuo

A Chinese Medical text, Benco Gangmu Materia Medica, by Li Shizhen (1578 AD), describes the use of cannabis in treating vomiting, parasitic infections, hemorrhage, diarrhea and dysentery and to stimulate the appetite. (Janet Joy PhD and Allison Mack, "Marijuana as Medicine: Beyond the Controversy," 2001).

Cannabis (hemp) spread to North and South America with European settlers; the Spaniards brought the plant to Chile in the mid 1500's. There is evidence that the Spaniards also brought cannabis to Florida -- perhaps even before Columbus. According to some historical records, American colonies preferred Spanish Hemp from southwestern Florida to hemp brought by Columbus. Even in early North American history, cannabis continued to be selectively bred to encourage higher THC levels and bushier plant characteristics and taller thicker hemp stems used for fiber.

Marijuana is a more recent slang word for a high THC content cannabis. The scientific name for higher THC Cannabis is sativa/indica. The word marijuana originated near the Mexican border when marijuana crept into the United States in the late 1930's. U.S. Oxford scholar Robert Burton recommended cannabis to treat depression. Source: National Institute of Drug Abuse (NIDA) Marijuana Research Findings 1976-77.

Marijuana was officially legal in the United States until 1937. Harry Anslinger, the first U.S. drug Czar, pushed through Congress the Marijuana Tax Act of 1937 -- with help from William Randolf Hearst, owner of a huge chain of newspapers. Hearst was heavily invested in the timber industry to support his newspaper chain and didn't want to see the development of hemp paper as competition.

The act was also supported by the Dupont chemical company and various pharmaceutical companies in the effort to outlaw cannabis. Dupont had recently developed and patented nylon and wanted hemp removed as competition in the manufacture of clothing and ropes. The Mical Company and various other pharmaceutical companies joined in the effort to outlaw cannabis. The reason? Pharmaceutical companies could neither identify nor standardize cannabis dosages; and with cannabis, folks could grow their own medicine and therefore not have to purchase it from large corporations.

On August 2, 1937, marijuana became illegal at the federal level with the passage of the Stamp Act. The law was signed by President Roosevelt and was designed to curb marijuana traffic through heavy taxes on transactions. This law created a federal ban on all marijuana possession--unless you had a stamp from the Treasury. The Treasury didn't make stamps, so it was thereby illegal. The United States feared that, after so much trouble had been caused by morphine addiction, marijuana would also be a gateway drug.

In 1964 Dr. Raphael Mechoulam, Professor of Medicinal Chemistry at Hebrew University in Jerusalem, was the first to identify delta 9 tetrahydrocannabinol(THC) in Cannabis, the compound that makes the marijuana smoker feel euphoria (THC is a psychoactive compound). He was the first to actually synthesize THC.

Since then, at least 60 different kinds of caniboids (CBD's) and THC's have been identified.

It could be speculated that cannabis evolved as a source of food, fiber and medicine. This is evidenced by the existence of many cannabinoid receptors found in the human body and likewise the many cannabinoids found in the seeds and leaves of cannabis. Again, cannabis must have been around for along time to allow this symbiotic relationship to evolve in humans and to keep the body DNA in balance. Present day research has also found 60 or more different phytocannabinoid molecules in marijuana and hemp plants and similar number of endo-cannabinoid receptors in the human body. The right combination of cannabinoids can treat diseases such as Dravet syndrome, a severe form of epilepsy, even more effectively than present patented pharmaceutical drugs on the market today.

Courtesy www.theroc.us

Again, cannabis is nothing new as a medical or psychological aid for the health and well being of patients worldwide, past or present. Just like salmon coming back to their historic streams to spawn and thrive, marijuana is coming back to civilizations worldwide to provide improved health, both physically and mentally.

Marijuana and hemp consist of both female and male plants. Marijuana is collected from the flowering buds of the female plant. These buds, harvested at the right time, have the optimum content of THC. There are many different varieties of Marijuana to fit the users liking, either by smoking, vaporizing or consuming extracted oils or mixing cannabinoid oils with food or candy. Marijuana has continued to provide healing properties to the present day.

Marijuana was listed in the United States Pharmacopeia from 1850 until 1942 and was prescribed for various health problems. Its use as an intoxicant was also commonplace from the 1850's to the 1930's. A campaign conducted in the 1930's by the U.S. Federal Bureau of Narcotics (now the Bureau of Narcotics and Dangerous Drugs) sought to portray marijuana as a powerful, addictive substance that would lead users to narcotic addiction. It is still considered a "gateway" drug by some authorities.

In June 1971 Nixon declared the war on drugs which took the drug policy of the United States worldwide. Nixon and his supporters labeled illegal drugs as public enemy number one in the United States. Nixon also created the DEA and increased penalties for marijuana. In September 1969 Nixon essentially shut down the US-Mexico border with "Operation Intercept" in an attempt to stop smuggling marijuana from Mexico to the US. Everyvehicle crossing the boarder was given a 3 minute intensive search. The operation lasted only 20 days after border delays severely affected the economies of both countries. The amount of drugs seized was much less than Nixon expected but that was not the true purpose for the unilateral action. It was to coerce Mexico into funding a domestic drug eradication and enforcement policy in which Operation Intercept was a success.

Operation Intercept was the first drug related unilateral action against another country bythe United States. It symbolized the US 's inability to accept responsibility for it's own drug problem of black market distribution within the US. In 1970 the Nixon government passed the Comprehensive Drug Abuse Prevention and Control Act which included the Controlled Substances Act

(CSA). The CSA created the five schedules which are used today in compliance to the UN Single Convention on Narcotic Drugs. The CSA increased the search and seizure powers of law enforcement and established a major federal scientific inquiry called the National Commission on Marihuana and Drug Abuse (aka the Shafer Commission). Marijuana was placed on Schedule 1 controlled substance which is the most restrictive. Crack cocaine and crystal meth are listed under Schedule 2. Marijuana's inclusion in Schedule 1 was to be temporary pending the results of the Shafer commission. When the commission released its findings to President Nixon in 1972 he was not pleased. The Shafer report found, "Neither the marijuana user nor the drug itself can be said to constitute a danger to public safety." The report also found that marijuana use caused "no significant physical, biochemical, or mental abnormalities," and that "most users, demonstrate an average or above-average degree of social functioning, academic achievement, and job performance." The Shafer commission also recommended that marijuana be decriminalized and at least be removed from its Schedule 1 status. Nixon refused to even read the report as revealed in a press conference."

American President Richard Nixon 1969-74
© Dawn Hudson | Dreamstime.com

In March 1973 Nixon created the Drug Enforcement Agency. The new "super agency" was consolidated from the Federal Bureau of Narcotics and Dangerous Drugs (BNDD), Customs, the CIA and the Office of Drug Abuse Law Enforcement (ODALE). Demand for illegal drugs has climbed uninterrupted since 1973, prices have not gone up and criminal gangs reap higher profits than ever. The DEA budget has increased annually from $75 million in 1973 to $2.5 billion in 2009. http://www.justice.gov/dea/agency/staffing

After 1939, most marijuana came from Mexico, but in 1975 the Mexican government agreed to eradicate the crop by spraying with the herbicide parquet, raising fears of toxic side effects to the workers and consumers. Colombia then became a main supplier of marijuana. The "zero tolerance" of the Reagan and Bush administrations (Just Say No!) resulted in the passage of strict laws and mandatory sentences for possession of marijuana and also strengthened forces at the southern borders.

Domestic cultivation of marijuana, particularly in Hawaii and California, continued to grow. Beginning in 1982, the Drug Enforcement Administration turned attention to busting marijuana farms in the United States, and there was a shift by growers to the less apparent indoor marijuana operations; this event created a new type of plant focus, specifically for small size and high yielding plants. After a decade of decreasing use, marijuana smoking began an upward trend once more in the early 1990's, especially among teenagers. http://www.narconon.org/drug-information/marijuana-history.html

The DEA finally held public hearings on the issue before an administrative law judge. Two years later [on Sept. 6, 1998], Judge Francis Youngruled [in the matter of Marijuana Rescheduling Petition, Docket No. 86-22] that the therapeutic use of marijuana was recognized by a respected minority of the medical community, and that it met the standards for alternative legal medications.

Prior to the [Nov. 3, 1998] election, former Presidents Ford, Carter, and Bush released a statement urging voters to reject state medical marijuana initiatives because they circumvented the standard process by which the Food and Drug Administration (FDA) tests medicines for safety and effectiveness.

"Compassionate medicine," these leaders insisted, "must be based on science, not political appeals." Nevertheless, medical marijuana initiatives proceeded to pass in most every state in which they appeared on the ballot.

The US Department of Health and Human Services received a patent (US 6,630,507 B1), Oct. 3, 2007, for the therapeutic use of 'cannabinoids as antioxidants and neuroprotectants.' The abstract says in part: "Cannabinoids have been found to have antioxidant properties . . . and have particular application as neuroprotectants . . . and in the treatment of neurodegenerative diseases such as Alzheimer's disease..." (Source: Janet Joy, PhD and Alison Mack, Marijuana as Medicine: Beyond the Controversy, 2001, National Organization for the Reform of Marijuana Laws (NORML) "Active State Medical Marijuana Programs," www.norml.org)

California became the first State to legalize Marijuana for medical use in November, 1996, followed by Alaska, Oregon, and Washington in 1998. These laws removed state level criminal penalties on the use, possession and cultivation of marijuana by patients who possess written documentation from their physician stating that they might benefit from the medical use of marijuana. Other states to legalize marijuana for medical use followed in order: Maine 1999, Hawaii 2000, Colorado 2000, Nevada 2000, Montana 2004, Vermont 2004, Rhode Island 2006, New Mexico 2007, and Michigan, the 13th state, passed the law in November of 2008, followed by New Jersey 2010, Arizona in 2010 and Delaware in May 2011. In the November election in 2012, Connecticut became the 17th state and Massachusetts the 18th to pass medical marijuana legislation. In July 2013 New Hampshire and Illinois became the 19th and 20th states to legalize medical marijuana. Washington DC officially passed a medical Marijuana law in 2010 but started dispensing on August 1, 2013. Maryland, Minnesota and New York, the 23rd state to pass medical marijuana legislation in 2014. Also in2014, nine more states, Wisconsin, Utah, South Carolina, Missouri, Mississippi, Kentucky, Iowaand Alabama, and Florida passed bills that allow the limited use of cannabis-extract oil for people with severe epilepsy, though many of those states do not allow personal growing of medical marijuana.

On July 22, 2010, the US Department of Veterans Affairs released a Veterans Health Administration (VHA) directive saying that veterans who participate in legal state medical marijuana programs will no longer be disqualified from "substance abuse programs, pain control programs, or other clinical programs." "VA Directive 2010-035" (15 KB), Department of Veterans Affairs website, July 22, 2010.

On June 22, 2011, a bipartisan team of US House of Representatives lead by Barney Frank, D-Mass., and Ron Paul, R-Texas, introduced the bill "Ending Federal Marijuana Prohibition Act of 2011." Other sponsors of the Frank-Paul legislation included Reps. John Conyers, D-Mich., Steve Cohen, D-Tenn., JaredPolis, D -Colo., and Barbara Lee, D-Calif. This proposed legislation would permit states to legalize, regulate, tax and control recreational marijuana without federal interference. The bill would limit the U.S. government's role in marijuana enforcement to only interdiction of cross-border or inter-state smuggling. Citizens would be able to legally grow, use and sell cannabis in states which have legalized medical marijuana and move the U.S. towards decriminalization and an education and treatment-based drug policy. The legislation also followed a report by the Global Commission on Drug Policy, released in June 2011, which pronounced the War on Drugs a failure and advocated the legal regulation of marijuana. The bill was not passed and the federal government continues to declare marijuana as a Schedule 1 Controlled Substance.

The governors of Washington and Rhode Island petitioned the US Drug Enforcement Agency (DEA) to reclassify marijuana from the most restrictive Schedule I category to a Schedule II category which, if approved, couldlead to pharmacies and other private stores dispensing marijuana. Schedule I Controlled Substances include LSD, Ecstasy, Peyote, Heroin, and other more abusive drugs. The 106 page petition, filed on November 30, 2011 byDemocratic Governor Christine Gregoire of Washington State and Independent Governor Lincoln Chafee of Rhode Island, declared that the Schedule I classification of marijuana is "fundamentally wrong and should be changed." Schedule II drugs, considered less abusive than schedule 1 drugs, include Opium, Cocaine, Methamphetamine, Demerol, Oxycontin, Adderall, Ritalin and others.

The failure of marijuana prohibition and its inherent hypocrisy are certainly becoming clearer than ever these days, and the discussion is starting to include some pretty big names. The report, released by the Global Commission on Drug Policy in early June 2011, is evidence that the end of The War on Drugs is going mainstream. The last three U.S. presidents have admitted to using marijuana. On June 16, 2011, Jimmy Carter, former president and life-long humanitarian, wrote in the New York Times that "the time has come to end the war on drugs." He suggested following the advice of the Global Commission and their recommendation to end marijuana prohibition. "Let's hope that some leaders who are still not convinced listen to President Carter and the Global Commission on Drug Policy. If not, the rest of us are prepared to keep shouting about the necessity of marijuana reform until laws are passed."

(Morgan Fox June 17, 2011). http://blog.mpp.org/prohibition/president-jimmy-carter-calls-for-end-to-war-on-drugs, 06/17/2011.

"Governors Asking U.S. to Ease Rules on Marijuana to Allow for Its Medical Use," New York Times, Nov. 30, 2011. Marijuana PolicyProject (MPP), Michael Cooper.

In the eight months following October 11th, 2011, the feds shut down at least 500 marijuana dispensaries in California, threatening criminal charges and the seizure of assets. The feds charged the dispensaries with violating not only federal law, but also state law, which requires operators to be primary caregivers to their customers and to distribute marijuana for medical purposes only. In other words, the dispensaries allegedly were also selling marijuana for recreational use on the side to make their operations more profitable. The crackdown, some have said, has made it difficult for qualified Californians to obtain marijuana for medical use, pushing medical patients into the black market.

In the November 2012 elections in both Colorado, Amendment 64, and in the State of Washington, Initiative 502, passed laws allowing adults over the age of 21 to possess and use marijuana for recreational purposes. Colorado Amendment A-64 allows adults 21 or older to grow up to three immature and three mature marijuana plants per household in a private, locked space. Adults can legally possess all cannabis from the plants they grow as long as they stay where it is grown. Adults can legally possess up to one ounce of cannabis while traveling, and give as a gift up to one ounce to other citizens 21 years of age or older. Consumption of cannabis or marijuana is permitted in a manner similar to alcohol, with equivalent offenses prescribed for driving under the influence. The amendment also provides for licensing of commercial marijuana cultivation facilities, testing facilities, and retail stores. The new legislation does not apply to medical marijuana for which rules were previously passed in the November election of 2000. The amendment also declares that industrial hemp should be regulated separately from marijuana, as hemp has much lower THC concentrations. Industrial hemp or any part thereof, shall not exceed a delta-9- tetrahydrocannabinol (THC) concentration of three tenths percent (0.3%) on a dry weight basis. It also requires the Colorado general assembly to "enact legislation governing the cultivation, processing, and sale of industrial hemp" by no later than July 1, 2014. This amendment also requires the general assembly to enact an excise tax to be levied upon wholesale sales of cannabis, requiring that the first $40 million in revenue raised annually by such tax be credited to the public school capital construction assistance fund.

The state of Washington's passage of Initiative 502 in the November 2012 general election removes state criminal and civil penalties for activities that it authorizes, including producing, processing, and selling marijuana. It is subject to licensing and regulation by the Liquor Control Board (LCB). It does away with state criminal penalties for adults 21 years or older in possession of "one ounce of dried marijuana or anyone growing marijuana without a license. The State of Washington imposes a 25% excise tax on wholesale and retail sales of marijuana. The city of Seattle is also considering an additional tax on marijuana sales. Washington's taxes from recreational marijuana are earmarked for substance abuse, prevention, research, education, and healthcare programs.

Again, the DEA is still keeping up big roadblocks by continuing to classify marijuana as a Schedule 1 Controlled Substance, along with heroin, LSD, Ecstasy and Peyote instead of classifying it a Schedule 2 Controlled Substance, along with Cocaine, Opium, Ritalin, Methamphetamine, Oxycontin and others. Under the Controlled Substances Act, any bank that takes money from marijuana commerce is technically illegal. An FDIC bank taking money from an illegal enterprise can lose its license and potentially be prosecuted for handling anything other than cash. Thus all medical and/or state legal marijuana enterprises have to deal with banks in cash, which is cumbersome to say the least. It further creates an underground economy with tax avoidance and criminal implications. A marijuana retail business can accumulate thousands of dollars a week in cash which exposes their business to criminal activity. Federal and State governments are working with banks to modernize bank transactions for marijuana and hemp commerce. Accepting credit cards and checks for marijuana transactions will also help vendors to deal with state and local tax authorities.

On February 14th 2014, The U.S. Treasury, with the approval of the Obama Administration, said banks can do business with marijuana enterprises in all forms of transactions in states where medical marijuana has been legalized and in states that have legalized recreational marijuana. The seven page Treasury Department edict indicated that it will not target bank transactions with marijuana businesses in those states where the businesses are properly licensed and their businesses don't show signs of gang or cartel-related activity. Some banks are still not going ahead with marijuana transactions until marijuana is officially moved from the Schedule 1 to a Schedule 2 drug classification. President Obama said in passed interviews that he views alcohol as a potentially more dangerous drug than marijuana. Recent statistics about the dangers of alcohol confirm Obama's statement.

Also in mid December 2014 the Obama Administration, through the Justice Dept. ruled that all Native American Tribes can create and enforce their own recreational marijuana laws, regardless of state laws. Each individual tribe has been given it's own authority to allow members 21 years or older to grow recreational marijuana for their own use or set up marijuana sales outlets. Each tribe can totally ban recreational marijuana on their tribal lands. Many tribes have been opposed to marijuana legalization. The federal government can continue to enforce prohibition for those tribes, at their request, even in states where pot is legal.

12 states have legalized Recreational Marijuana

Alaska, California, Colorado, District of Colombia, Illinois, Maine, Massachusetts, Michigan, Nevada, Oregon, Vermont, and Washington

46 States have legalized Medical Marijuana in some form.

Alabama, Alaska, Arizona, Arkansas, California, Colorado, Connecticut, Delaware, District of Colombia, Florida, Georgia, Hawaii, Idaho, Illinois, Indiana, Iowa, Kentucky, Louisiana, Maine, Maryland, Massachusetts, Michigan, Minnesota, Montana, Nevada, New Hampshire, New Jersey, New Mexico, New York, North Dakota, Ohio, Oklahoma, Oregon, Pennsylvania, Puerto Rico, Rhode Island, South Carolina, Tennessee, Texas, Utah, Vermont, Virginia, Washington, West Virginia, Wisconsin and Wyoming.

Cannabis Medicine has been sold in the United States for over 100 years. *Courtesy: www.blazingtrailsbend.com*

Old Cannabis Tinctures

http://antiquecannabilbook.com/chap4/Tincture.htm

In 1839, Dr. William O'Shaughnessy, published "On The Preparations of the Indian Hemp, or Gunjah." He was a young Irish physician residing in India. He had learned about Cannabis preparations from Mohammedan and Hindu physicians. Upon returning to England, he quickly went about lecturing about medical uses, but also lectured about his pharmaceutical preparations and manufacturing techniques.

Tinctures were already in use by Western Medicine for centuries before but O'Shaughnessy used Cannabis in his preparations. Cannabis Tinctures first began to make their appearance around 1840, into Brand/Trade name products. Also apothecary shops began to manufacture their own tincture up to the early 1900's. Cannabis tinctures were prescribed by doctors directly for numerous ailments. Usually the bottles recommended so many drops of Cannabis Tincture to be diluted in a quarter cup of warm water. Later most Tinctures were NOT being manufactured by established pharmaceutical companies, but by local apothecary shops throughout the country. Industrial hemp was commonly and legally grown up into the early 20th Century. Then established Pharmaceuticals started to take over with their brand names. Above are some classic tincture bottles and related rubs.

Charlotte'sWeb

By - Fred Gardner - O'Shaughnessy's News Service and Bruce Meland

There is a point at which the Charlotte's Web™ story goes from being a matter of interest to a fraction of the public to one that everyone's talking about, inescapable in the media. That point was reached on Sunday August 11, 2013 when Dr. Sanjay Gupta, MD Correspondent for CNN, narrated an hour-long special report titled "Weed", stating Charlotte's Web™, a unique strain of marijuana, provided dramatic examples of Cannabis exerting beneficial effects. Most memorable was the return to health of a five year old Colorado Springs girl, Charlotte Figi afflicted with Dravet Syndrome, a severe form of epilepsy.

Charlotte with high CBD-rich Charlotte's Web™ Cannabis. Courtesy of www.theroc.us

Charlotte's parents told Gupta how her condition worsened as every conventional treatment failed. She was having hundreds of seizures a day and the doctors were considering inducing a coma to save her life. Her dad, a Special Forces soldier deployed to Afghanistan, researched epilepsy treatments on the web and learned about CBD strains of marijuana. Desperate, the Figis bought $800 worth of CBD-rich cannabis at a dispensary and a friend helped extract the oil for under-the-tongue application. Charlotte's seizures went down to one a week and she began recovering! Soon there after the treatment of Charlotte Figi with Joel Stanley and The Realm of Caring Foundation, now called Stanley Brothers Social Enterprise, began.

Stanley Brothers Social Enterprise has given Charlotte a chance at life again. It has changed everything for her, her twin sister and for her parents. I try not to think about what our lives would look like had we not met the Stanleys and begun CBD treatment 2 years ago. "I feel certain Charlotte would not be here with us. I will be forever grateful for the compassion of the Stanley Brothers" said Paige Figi, mother of Charlotte Figi.

A Colorado grower named Joel Stanley and his brothers supplied the Figis with Cannabis tincture made from their newly developed strain of marijuana plant with a CBD-to-THC ratio greater than 20-to-1, and named their strain Charlotte's Web™. Parents of children with severe epilepsy began contacting the Stanley brothers with urgent requests for the CBD-rich oil. They created a new company, Stanley Brothers" Social Enterprise to facilitate distribution. Paige Figi, Charlotte's mother, works for SBSE as a volunteer.

"In early 2014 there were 179 children and 27 adults on Charlotte's Web™, according to Heather Jackson, RoC's executive director. "There are now over 3000 clients on CW Botanicals™," she adds. Stanley Brothers Social Enterprise does not charge for their services."

Desperate parents have been moving to Colorado to expedite access to CBD-rich cannabis for their kids. In many states, politicians, activists and parents are pushing legislation to legalize CBD rich cannabis (while maintaining the prohibition of THC). The Stanley brothers have testified in support of CBD-only bills with minimum THC in Florida, Alabama, Kentucky, Nebraska, Utah, New York, California, Georgia, Oklahoma, Maryland, Vermont, Wisconsin, Pennsylvania, and Illinois.

Among those who contacted the Stanley brothers in 2013 was Ray Mirzabegian, the father of Dravet syndrome patient, Emily, who was nine years old. After many trips to Colorado, Mirzabegian was provided with Charlotte's Web™ developed by The Stanley Brothers over the course of several years. Ray was also authorized to launch Realm of Caring California, a subsidiary of The Stanley Brothers Social Enterprise, created in Colorado. Mirzabegian started growing Charlotte's Web plants as per the Stanleys' specifications, and arranging to distribute the CBD-rich oil that was extracted from Charlotte's Web™.

On Feb. 23, 2014, in Milpitas, California he updated 120 parents of pediatric epilepsy clients on the state of his grow. Many in the Milpitas audience were on the waiting list for CBD-rich oil from the Charlotte's Web™ plants that Mirzabegian was growing at three locations in Southern California. "No more than ninety-nine plants per facility," he explained. "We are doing everything legally." He gave up a career teaching opticians to grow medical cannabis for his daughter Emily, now 10, and others in urgent need. His brothers Alfred, a general contractor, and John, a hair-transplant technician, are also spending 14-15 hours a day growing plants--organically--for RoC California.

Ray Mirzabegian seemed completely sincere and likable. He is 42, grew up and lives in the Los Angeles area; speaks almost perfect English with an Armenian accent. Here is a partial transcript of Ray's presentation:

"...because of her uncontrolled seizures — despite numerous medications and a ton of other treatments that failed — about 2 years ago after I watched the Stanley brothers in a documentary "American Weed" on CNN, I traveled to Colorado several times... and I was able to obtain CW Botanicals™ for my daughter and for everybody else."

I went into an agreement with the Stanley Bros. Social Enterprises to start growing plants here and offering oil to patients in California who needed it. It took about three months to realize how severely California needed this CBD oil. The waiting list was about 400 plus at that time, but also patients with other severe problems. I haven't really advertised it — just word of mouth and Facebook groups.

I'm working very hard to establish what is required legally and to learn how to do it right. I don't come from a background of growing. I'm a professor. I concluded my career about a month ago. I used to teach opticians in college for 20 years. But it took about a year or so to learn the science of growing this plant the right way — meaning to make sure that it's organic, that it's clean, that there are no pesticides, molds, anything else. To make sure that it's a quality product, because after all, I'm going to give it to my daughter.

I wanted this oil because my daughter was having many, many seizures. She had tried 13 medications that did not work. The ketogenic diet, an anti seizure diet low-in-carbs, high-in-fats, worked for about six months and then the seizures slowly started coming back. About a year after trying the diet, my wife and I sat down and decided that she has absolutely no quality of life. She was depressed because she couldn't eat anything at school or at home. We could have tweaked it more but we thought it's not really working. So we decided to stop the ketogenic diet and sure enough the seizures came back.

Charlotte's Web™ growing in greenhouse by Stanley Brother's Realm of Caring Foundation in Colorado Springs, CO.
Courtesy of www.theroc.us

At a very well known epilepsy center in Southern California, they casually pulled us aside and said, "There's really nothing else we can do for Emily, we suggest that you go home and enjoy your daughter as best you can." So we went home very disappointed and upset. We saw about eight or nine neurologists after that and then finally found one at UCLA who was willing to sit down and listen to us, to have a conversation. Our neurologist is very supportive of the parent's right to try something like CBD oil - especially since we've tried everything else. I talk to him frequently and he's interested in the feedback about the patients trying CBDs, and about my daughter. He's also supporting me in tapering off the medications because they're not working and all we're getting is negative side effects. If a medication is helping, I can understand putting up with the side effects. But if there's no seizure control... For instance, with my daughter's speech fluency: on the drug Depakote, it was like someone pushed the mute button on her.

I watched Charlotte's story on CNN, and I went back to Colorado and then I came back and said to myself

Leaf of Charlotte's Web™
Courtesy of www.theroc.us

I'm going to start a waiting list and a Facebook group to show these guys that there's a demand for this product in California. I emailed one of the Stanley brothers and he answered the next day. Then I went to Colorado. He said "We get about a thousand emails every week and we forward them to the foundation but I decided to answer yours, I don't know why, and here you are." I brought back a limited quantity of Charlotte's Web™. So I had to grow and grow and grow and clone and clone and clone because I had 400 patients waiting for it. I have a feeling that number is going to increase after the second documentary on CNN with Dr. Sanjay Gupta.

The leaf of Charlotte's Web™ appears to be streaked with yellow and some pink — variegated in color. It's a beautiful plant and very different from everything else I've seen. It's a very fragile plant. It requires a lot of love and singing and care taking.

My daughter is speaking more and more ever since we started her on Charlotte's Web™ hemp oil and were able to take her off most of the medicine. Emily is almost free of pharmaceutical medication now (applause!). At times she's been on four different meds and while on those she's had several seizures a week. She's now eating everything on her own, socializing, running, jumping, you name it. And since tapering off the last Benzo medication and Clozapine--which has done so much damage in the past—Emily seems to feel so much better. Even as we were tapering off this medication, she was averaging one seizureevery ten days.

I'm hopeful that once we're completely off the pharmaceutical meds and dial in the right dose of the hemp oil, the right ratio, that she can have the same success as Charlotte Figi and everyone else who is seizure--and prescription drug-free. There are a couple of things wrong with my daughter. One was the grand epilepsy, but the side effects of the pharmaceuticals were even worse.

I talk about this with her neurologist. We send each other articles and research papers. We communicate. I finally found a good one who listens, who researches. And I suggest you do the same. And I suggest that you do not touch this medication unless you have that doctor who is supporting you because these withdrawals can be crazy and they're sometimes scary. But if you have the right doctor to guide you through it, you will understand that what you're going through is okay and will pass and that you can call the doctor and ask, "Should I go to the hospital or not?" We haven't been to any hospital for three years now. There are a lot of success stories out there. Not just from Realm of Caring but many companies that are also offering marijuana as a medicine. It's a one-on-one personal thing. Some people do okay, some people do great, and some people need more time to find the right dose.

And that's something that we all need to understand. We have to have discussions within our groups: What times of day? How many milliliters? Before or after meals? Try to document everything. One of the biggest things that helped my daughter was that, at 4:30 in the morning, we gave her half a milliliter of CBD while she was sleeping. That was such a big change in her! It almost completely stopped her seizures that happened in the early morning. These are the things you have to understand: you have to try it; you have to give it time; you have to document everything.

There's a lot of discussion going on about THCA and THC. Some patients will do well on it, but some will do better on CBD only. It's so hard for me to tell all 400 people to just hang in there and be patient. I wish there was something I could do to make these plants grow faster — but I want to grow organically with no extra hormones. So, we're almost there. If all goes well, I should have three harvests in 2014. And between those three harvests I'm hoping that everybody will get their medicine.

Charlotte is doing great. She went from hundreds of seizures a week to almost none and is now pharmaceutical -free. I just met another patient, Zaki, in Colorado, who is doing very well. Zaki is the son of Realm of Caring Director, Heather Jackson. You would not know that he comes from a past of seizures and other issues. He talks, communicates, runs and plays like the other kids. We were sitting there talking about football and some of the players and I thought, if my daughter could get this far.... It was such an inspiration! Zaki has been seizure free for over a year. My daughter is happy and active and doing much better than she was on all those prescription medications. But it's a work in progress. I can't wait until the plants are mature and the medicine is ready so I can offer it to all these children--and then hear all of your success stories. That will be the best day of my life--and it's coming soon!

I want to clarify something: I'm not here to say that Charlotte's Web™ is the only CBD oil that works. On the contrary, there are a lot of great plants out there. There are some companies that are doing good things and I'm supporting them. I have sent many patients to companies that are trying to offer viable treatments. ACDC is a great strain. But it must be grown in the right way to insure that it's stable and clean. There are more companies coming with CBD oil and I'm happy about that. I wish more companies would come up and pay attention to our children. So we will support each other and work together. I will be very happy when this happens!

They will continue to offer great services, educating and advocating, setting standards with the promise of the best quality product at low prices with a guaranteed supply.

http://www.dailynews.com/health/20140612/realm-of-caring-californiaraises-money-for-grow-house-after-relocatingin-la#article-topfundndraising.
Ray Mirzabegian www.theroc.org and www.beyondthc.com

Photo of Charlotte's Web™ CBD Oil and products. Courtesy of www.theroc.us

RICK SIMPSON SKIPS JAIL TIME IN CANADA FOR CURING CANCER - SEEKS ASYLUM IN EUROPE

RCMP destroys all but 1 marijuana plant

In 2003 Rick Simpson, a hands on pioneer for curing cancer in Nova Scotia, was diagnosed with basal cell carcinoma skin cancer. He had spots of cancer on his body, face and neck. He tried surgery with no results. Rick had been extracting oil from marijuana and taking it orally for general good health. He had heard of studies that cannabinoids from marijuana could kill cancer in mice. He decided to apply his oil via soaked bandages to his cancer sores. After 4 days he pulled the bandages and the cancer was gone! He told his doctor and cancer organizations but they would not listen. He decided to grow his own marijuana on a larger scale and make his own oil and give it away to friends and anyone who needed it. Despite having his garden raided by the RCMP he continued his operation. In 2008 a documentary was made of Rick's story titled "Race From The Cure". It can be viewed on YouTube, millions have watched it worldwide.

In over 10 years his hemp oil has cured thousands of cancer victims, and has spurred other individuals and companies to follow his work. On November 25, 2009, a day before he was to receive the Freedom Fighter Of The Year award at the Cannabis Cup in Amsterdam, he received word that his garden had been raided again by the RCMP. Rick believed the purpose of the raid was to keep him from coming home and continuing his good work. Also if he were to come home he would likely be put in jail; thereby denying production of his oil for his patients and continued good health. He decided to seek asylum in Europe and spread the good word about his specialty, Rick Simpson oil, at seminars around Europe.

Here is a summary, in Rick Simpson's own words: I no longer live in Canada although, as far as I know, no police agencies are still looking for me. Canada is truly a beautiful country and there are good people within its boundaries. Due to my criminal record in Canada I cannot even live in the U.S. At present we have many governments in the Balkans cornered; and they will have to completely legalize the medicinal use of my oil. There were also some other personal reasons I decided to stay in Europe.

On countless occasions I have told the public I don't produce or supply my oil, as it is against the law in many countries. I am not connected with anyone who claims they are selling "The Real Rick Simpson Oil". It is impossible to stop all these suppliers. Now the public is aware, so please be aware of this fact. I am not connected to any foundation, like the Phoenix Tears Foundation and Janet Sweeney. The public must realize there are no standards or quality control to know how to make my oil. To obtain my real oil you have to produce it from directions on my website. I hope these instructions will allow vast numbers of those who are suffering to find a sensible solution to their medical issues and inform others who are currently in need of this oil's healing powers. You can get all my information from my website www.phoenixtears.ca.

Best Wishes

Rick Simpson

Dr. Raphael Mechoulam: World's Foremost Cannabis Reseacher

Discovers THC and CBD Molecules, endocannabinoid receptors, Anandamide and CB1 and CB2 receptors in the Human Body

FIRST MAJOR BREAKTHROUGH: ISOLATION OF THC

Dr. Mechoulam was the first to isolate Δ9-tetrahydrocannabinol, known by the acronym THC. THC is the most abundant cannabinoid secreted in the trichomes of cannabis plants and is responsible for emitting psychoactive effects.

Mechoulam also successfully elucidated and isolated the second most abundant cannabinoid: **cannabidiol (CBD)**. CBD surprised Mechoulam with its apparent medical efficacy as an anti-psychotic and possible ingredient in the treatment of chronic pain and seizures.

Mechoulam's total synthesis of cannabinoids eventually led him to the discovery of the human endocannabinoid system and the close relationship between cannabinoids and human physiology. He found that cannabinoids bind to receptors in different areas of the brain and vital organs.

In 1992, he found that people naturally produce their own internal cannabinoids that maintain importance in regulating human mood, pain, memory and more. In his research, Mechoulam discovered the chemical anandamide. Anandamide is characterized as a fatty acid neurotransmitter and has been linked to helping suppress cancer cell proliferation, among other benefits.

ANANDAMIDE AND 2-AG

Anandamide's effects are mediated by two primary cannabinoid receptors, CB1 and CB2. CB1 interacts with anandamide in the central nervous system, while CB2 is responsible for outcomes associated with the peripheral nervous system. Followed by his discovery of anandamide, he found in 1995 that 2-AG was another neurotransmitter maintaining unique importance in human psychological and physical regulation. Mechoulam has linked cannabinoids to helping to suppress addiction and aiding with traumatic head injuries due to blood vessel activity.

Mechoulam received the distinct honor of the Israel Prize. He has been awarded multiple lifetime achievement awards and received the Rothschild Prize in 2012 for his discoveries on the influences of natural chemical substances on human behavior.

Dr. Mechoulam currently teaches at Hebrew University in Jerusalem. His research has been cited over 40,000 times and is considered to be the standard on which further scientific development will be based.

Photo credits: **Tzahy (Wikimedia Commons)**

https://www.cannabis.info/en/blog/dr-raphael-mechoulam-worlds-foremost-cannabis-reseacher

The first endogenous cannabinoids, Anandamide and 2-AG, were found in Dr. Mechoulam's laboratory in Israel in 1992.

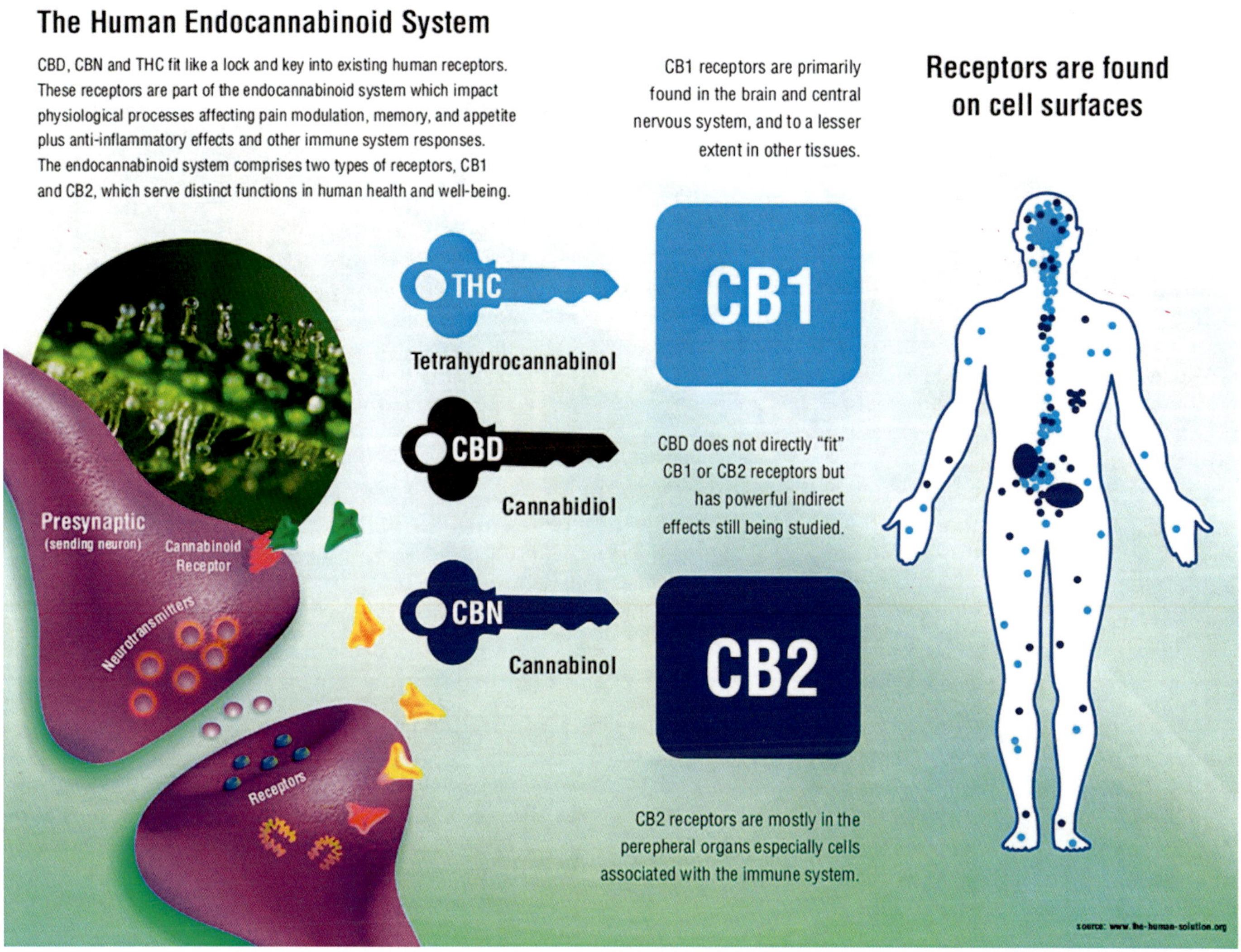

Dr. Lester Grinspoon, Marijuana Pioneer

Dr. Lester Grinspoon is Associate Professor Emeritus of Psychiatry at Harvard Medical School. Grinspoon was senior psychiatrist at the Massachusetts Mental Health Center in Boston for 40 years. Dr. Grinspoon is a fellow of the American Association for the Advancement of Science and the American Psychiatric Association. He was founding editor of the The American Psychiatric Association Annual Review and the Harvard Mental Health Letter. Grinspoon was editor of the Harvard Mental Health Letter for fifteen years. Dr. Grinspoon became inter ested in marijuana in the 1960s when its use in the United States increased dramatically. He "had no doubt that it was a very harmful drug that was unfortunately being used by more and more foolish young people who would not listen to or could not believe or understand the warnings about its dangers." When Grinspoon began studying marijuana in 1967, his intention was to "define scientifically the nature and degree of those dangers" but as he reviewed the existing literature on the subject Dr. Grinspoon reached the conclusion he and the general public had been misinformed and misled. "There was little empirical evidence to support my beliefs about the dangers of marijuana," and he was convinced cannabis was much less harmful than he had believed. His book Marihuana Reconsidered" reflected that change in view." He has testified before Congress, and as an expert witness in various legal proceedings, including the deportation hearings of John Lennon. Grinspoon worked with Ramsey Clark on a number of international marijuana related incidents. Grinspoon is widely known as the author or co-author of several cannabis/psychedelic-related books, including Marijuana Reconsidered.

Dr Lester Grinspoon www.norml.org

The photo of Dr Grinspoon was taken at the 2003 NORML (National Organization to Reform Marijuana Laws) annual convention at the Bay Front Hyatt Regency hotel in San Francisco, April 20 (4/20) Dignitaries, Doctors, Politicians, Activists attended. Efforts continued to place marijuana in the positive light. Dr Grinspoon talked about pro and cons various methods of using marijuana.

PROS AND CONS of HEALTH ISSUES OF TEENS USING MARIJUANA

UC San Diego researchers scanned the brains of 92 young people, ages 16 to 20 over a 18 month period. About half of participants with history of alcohol and marijuana use continued drinking. The other half abstained or maintained minimal use. The before and after scans showed that teens drinking at least 5 drinks per week had reduced brain tissue health. The brain tissue of teens who used only marijuana up to 9 times per week showed no physical damage. Alcohol use "becomes a cycle," said Joahana Jacobus of UC San Diego.

Teens who drink regularly damage areas of the brain that influence self-control and judgment. Some marijuana may not be as harmful in part because lower THC strains may actually have neuro protective effects. The study observed physical brain tissue, not overall performance of the participants.

In a 2011 study The American Academy of Child Adolescent Psychiatry found teen marijuana use is at its highest in 30 years and teens are now more likely to use marijuana than tobacco. The Academy found that parents can play an important role in teaching their children about the risks and consequences of using marijuana. The Academy website lists subjects to teach their children about the dangers of marijuana and signs of marijuana use. Many teenagers believe that marijuana is safer than alcohol or other drugs. The study goes into detail on the effects of marijuana use by adolescents and teenagers. The study concludes marijuana use in teens can lead to long-term problems.

According to a study Published in the April 16, 2014 issue of The Journal of Neuroscience: The size and shape of two brain regions involved in emotion and motivation may differ in teens and young adults who smoke marijuana at least once a week. These regions of the brain are associated motivation, attention, learning and memory. Studies have shown high THC marijuana exposure can causes structural changes in these brain regions. However, less is known about how low to moderate marijuana use affect brain structure. Researchers at Northwestern University and Massachusetts General Hospital/ Harvard Medical School used MRI imaging to compare the brains of 18- to 25-year old's who reported smoking marijuana at least once per week with those with little to no history of marijuana use. MRI data revealed significant brain differences. The nucleus accumbens -- a brain region known to be involved in reward processing -- was larger and altered in its shape and structure in the marijuana users compared to nonusers.

http://cir.ca/news/marijuana-teen-brain-tissue-study

In first pot-legal states, Colorado and Washington, teen use of marijuana with ultra high content of THC (25% or more) has resulted in more frequent emergency visits for treatment of vomiting, paranoia and psychosis. Source: Jennifer Oldham, Special to The Washington Post, The Bend Bulletin, June 16, 2019.

Brief History Of William O'Shaugnessy A Marijuana Pioneer

In the 19th century, Irish physician William Brooke O'Shaughnessy documented preparations of hemp, or gunjah, during his time in India working for the East India Company. O'Shaughnessy was a surgeon, physician, and professor of chemistry, a scientist who was an expert in botanical pharmacology.

According to "The Remarkable W. B. O'Shaughnessy," an essay written by Dr. Michael R. Aldrich, little was actually known about cannabis therapy in Europe and America until O'Shaughnessy read a paper to a group of students and scholars of the Medical and Physical Society of Calcutta in 1839. The 40-page paper was a model of modern pharmaceutical research. It included a thorough review of the history of cannabis for medical uses by Ayurvedic and Persian physicians in India and the Middle East. Aldrich's essay was published in The O'Shaughnessy's Reader, an ongoing online history of the medical marijuana movement found at www.beyondthc.com. Aldrich wrote that O'Shaughnessy also conducted clinical trials of cannabis preparations, first with experiments on mice, dogs, rabbits and cats, then by giving extracts and tinctures (of his own devising, based on native recipes) to some of his patients. O'Shaughnessy presented case studies of patients suffering from rheumatism, hydrophobia, cholera, rabies and tetanus, as well as a 40-day-old baby with convulsions, who responded well to cannabis therapy, leaping from near death to "the enjoyment of robust health" in a few days.

O'Shaughnessy took a furlough back to England in 1841. He brought quantities of hemp for the Pharmaceutical Society and specimens of cannabis indica and Nux vomica back to the Royal Botanical Gardens at Kew. His work

was finally published in 1964 (Gaoni & Mechoulam 1964). Even Sir J. Russell Reynolds, personal physician to Queen Victoria, recommended it to his patients for menstrual cramps. O'Shaughnessy was elected a Fellow of the Royal Society in 1843.

O'Shaughnessy's paper caused a sensation when it became widely available in England, Aldrich wrote. He had introduced a wonder drug to treat some of the most awful medical conditions of the 19th century. Physicians throughout Europe and America tried cannabis for a huge variety of illnesses. A similar thing happened when Dr. Tod Mikuriya reprinted O'Shaughnessy's paper as the lead article in "Marijuana: Medical Papers 1839-1972." It reinvigorated medical interest in the drug and sparked hundreds more articles on cannabis therapy into the 21st century. During the following years he also wrote on subjects relating to telegraphy, including a book of Private Codes for encrypted telegraphy. In 1860 he returned to Europe for sick leave where he remained in obscurity until his death in January 1889.

http://www.ccrmg.org/wbos/index.html

http://www.examiner.com/x-7002-Pittsburgh-Neighborhood-History-Examiner~y2009m6d27-William-B-OShaughnessy-hemp-gunjah-history-of-medical-marijuana-research-evidence-case-studies

---

The Prohibition Industrial Complex

by Sabrina Fendrick,

Director of Women's Outreach
Part of NORML.org

Our nation's marijuana laws are being held hostage by a prohibition industrial complex.

The latest Wall St. Journal/NBC poll shows, yet again, that the majority of Americans support legalizing the recreational use of marijuana for adults age 21 and over. But despite this surge in support (several other national polls have seen similar results), there are a few well financed, politically powerful groups that remain staunchly against reform – and which will likely serve as the biggest hindrance to widespread change. These folks have made a lot of money off of marijuana's current legal status, and those individuals (as well as their businesses/shareholders) are deeply invested in making sure things stay the way they are. The wide range of direct and auxiliary enforcement mechanisms, as well as the increase in drug testing laws, are driven by companies and businesses who provide the services necessary to support this disastrous and wasteful policy.

One such industry that has a financial interest in maintaining the status quo is law enforcement, especially drug officers and private prisons. Drug officers benefit from forfeiture and federal grants; private prisons keep their jails full and multi-million dollar state contracts in place. The Office of National Drug Control Policy requested $9.4 billion in funding for 2013, the majority of which went to enforcement and incarceration.

More specifically, California police –- one of the most vocal opponents to legalization in the state--made $181.4 million by seizing and selling the homes and cars of Californians involved in marijuana cases from 2002 to 2012. Police in Washington State are already taking budget hits as a result of the passage of I-502, the state's marijuana legalization initiative that passed in 2012. It was reported that some police drug task forces lost 15 percent of funding due to decreased revenue from marijuana forfeiture cases. On a national level, marijuana cases netted $1 billion in assets forfeited between 2002 and 2012. Assets can be seized under federal or state law, depending on the situation.

The Wall St. Journal recently reported that marijuana law reform would cut into a significant percentage of drug task forces' revenue. Most cash generated from drug-related propertyforfeitures goes to the law enforcement agency that made the bust. The Journal reports that, "Nationally, assets forfeited in marijuana cases from 2002 through 2012 accounted for $1 billion of the $6.5 billion from all drug busts." Task forces also rely heavily on federal grants.

One example of a federal grant relied heavily upon by drug task forces is Edward Byrne Memorial Justice Assistance Grant Program. The amount of money distributed is based on the number of drug arrests made for that year, among other components. The more drug arrests made, the more grant money provided--and 50% of all drug arrests are marijuana related. No drug will be able to fill the void of marijuana arrests. Marijuana is easier to spot and smell and is consumed by more people than any other illegal drug making marijuana arrest rates a significant percentage of overall revenue. And then there are the state contracts with private prisons, which mandate that facilities be filled at 90% capacity at all times. If 50% of inmates are there as a result of drug-related crimes, and half of that is for marijuana, then legalization would be a serious threat to new contracts and increased profits.

Another industry tied into the prohibition industrial complex is the drug testing market. It's a multi-billion dollar a year industry with its own built-in legislative advocacy machine. For example DATIA, the Drug & Alcohol Testing Industry Association, is the industry organization that represents more than 1,200 companies and employs a DC-based lobbying firm Washington Policy Associates. Their mission statement includes, among other things, creating "new opportunities for the drug testing industry."

In 2002, a representative from the influential drug-testing management firm Besinger, DuPont & Associates (Robert DuPont, Nixon's first drug czar is a high profile opponent to legalization) heralded schools as "potentially a much bigger market than the workplace." Work-place drug testing is a declining market due to the fact that employers see minimal return on investment. In fact, a DATIA newsletter dubbed school children "the next frontier." Not surprisingly, this industry advocates testing in all grades and for all extracurricular activities. It should be noted that several reports have concluded that drug testing minors is not only ineffective but can be emotionally and psychologically damaging. Fortunately, many schools have been reluctant to embrace testing.

Year after year, the drug testing industry gears up for another legislative push, ghostwriting bills for local and national lawmakers demanding testing for people who receive public assistance. Many of these elected officials are either financially invested in these companies or received significant financial contributions from industry organizations. For example, in February 2012, Congress amended federal rules to allow states to drug-test select job applicants. Among the lawmakers advocating for the change was Congressman Dave Camp, who owns at least $81,000 in assets in companies that are major players in the drug-testing industry, such as LabCorp and Abbott Laboratories. He has also received $5,000 in federal campaign contributions from LabCorp over the past three years. In the lead-up to the 2010 and 2012 elections, Abbott Laboratories spent $133,500 on campaign donations to Ohio and Texas state politicians promoting drug testing in welfare recipients--in addition to more than $500,000 spent by the company on state lobbying contracts since 2010. The industry is once again flexing its political muscles by pushing for policies that mandate drug testing for welfare recipients. Legislation has already been introduced in Virginia, New York, Arizona, Ohio, Iowa, Illinois and Mississippi for the 2014 legislative session.

Two of the most outspoken opponents of marijuana legalization are David Evans and Robert DuPont. DuPont, Founder of Besinger, DuPont & Associates, served as the nation's first drug policy director under Presidents Richard Nixon and Gerald Ford. During that time he had advocated decriminalizing marijuana and described its use a "minor problem." Once he left public office however, he became a "drug-testing management" consultant. David Evans worked for Hoffmann-La Roche, a multi- billion dollar drug testing group encouraging workplace drug testing policies. He now runs his own lobby firm and has ghost written several state laws to expand drug testing. Drug testing, overall, detects marijuana more than any other drug, which stays in the body for up to a month — as opposed to other harder drugs like cocaine and heroin, which are metabolized within one to three days. That is why they have such significant stake in keeping the plant illegal.

The total revenue for all of these industries combined adds up to hundreds of billions of dollars annually--a significant amount derived from taxpayer dollars. An industrial complex occurs when there is a policy and monetary relationship between legislators, the public sector and an industrial base that supports them. Just like the military industrial complex, the prohibition industrial complex, and its cycle of laws, enforcement and contracts poses a major challenge to reform efforts. This will be especially true in states that don't have ballot initiatives, which is why it is so important for everyone to become active on a local level and to hold lawmakers accountable. Though difficult, this is not an impossible challenge to overcome--as long as we remain diligent and active in the political process.

See more at: http://blog. norml.org/2014/01/28/the- prohibition-industrial-complex/#sthash. uAQaD-KVq.dpuf

Real Marijuana Found In Ancient Asian Tomb

By Bruce Meland

Farmers cultivating land in the Gobi Desert in Central Asia near Turpan China in the early 1990's discovered a large ancient cemetery of several acres. One of the unearthed 6 foot deep 2700 year grave revealed the skeletal remains 45 year old male of high social status commonly referred to as a Gushi Shaman. Artifacts supporting his noble status included a wooden bowl in a leather basket at both his head and feet. In one bowl discovered nearly 1½ pounds of marijuana leaves and buds.

Cannabis seeds were found amongst the cannabis in the bowl. Also found buried close to the skeleton were archery equipment, horse bridles, pots and a kongou harp and other materials supporting his shaman status. It was traditional to place items needed for the afterlife in the tomb with the departed. He apparently was a member of the Chinese-European nomadic clan as evidenced by his light hair and blue eyes.

All the details of the unusual preserved cannabis material can be found in a paper in the 2008 issue of the Journal of Experimental Botany. Lead Author of the study, Ethan Russo MD and former adjunct professor in the department of Pharmaceutical Sciences at the University of Montana and clinical professor in the

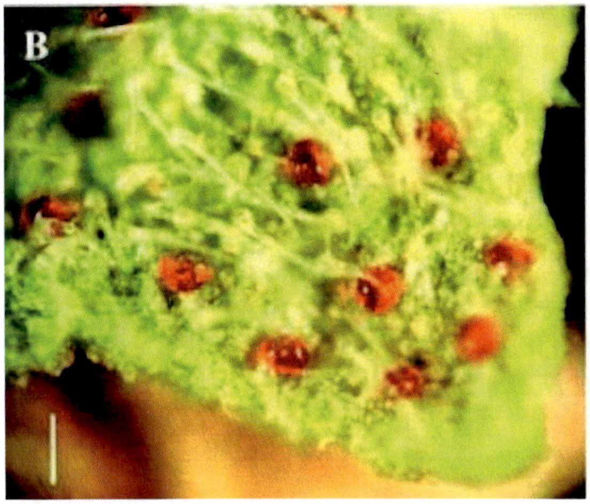

Photo micrograph of marijuana leaf fragment.
Photos Credit: David Potter/Oxford University Press
http//jxb.oxfordjournals.orgcontent/59/15/4171.full

Dept. of Medicine at the University of Washington revealed that the marijuana found in the shamans grave is quite similar" to what's grown today. We know from both the chemical analysis and genetics that the plant could produce THC (tetrahydrocannabinolic acid synthase, the main psychoactive chemical). Principal author, Dr. Ethan Russo added that no one could feel its effects today, due to decomposition over many centuries. Dr. Russo served as a visiting

Photo micrograph of ancient marijuana seed.

Photo of 2700 year old Indo-European Shaman grave. This photo was previously published in Mandarin (Xinjiang Institute of Cultural Relics and Archaeology, 2004 and used with permission.

Courtesy of Oxford Journals http://jxb.oxfordjournals.org/content/suppl/2008/10/03/ern260.DC1/00027516-file001.pdf

professor at the Chinese Academy of Sciences Institute of Botany while conducting the study along with his international team that also analyzed the ancient cannabis. Detailed chemical procedures were followed in analyzing the ancient cannabis including sterilization and handling with gloves in a laminar flow hood and metal spatulas. Microscopic observations and photos were taken to further define the samples. This is the first real evidence of high THC cannabis found in Central Asia. About ten thousand years ago hemp fabric was found in Taiwan. The ancient marijuana remnants and other burial artifacts are now housed at Turpan Museum in China. In the future, Russo hopes to conduct further research at the Yanghai site, which has 2,000 other tombs.

Sources: Phytochemical and Genetic Analysis of Ancient Cannabis from Central Asia Journal of Experimental Biology Vol. 59 Issue 15 p. 4171- 4182 11/01/2008 ©. By permission of Oxford University Press on behalf of the Society for Experimental Biology and Dr. Ethan Russo, Principal Author and Medical Director, PHYTECS.com

Recent Studies Show Medical Marijuana May Help NFL Brain Injuries

by Bruce Meland

Recent brain Injury treatment studies in Israel, Germany and the US show Medical Marijuana appears to be one of the best alternative treatments at present for brain injuries including concussions. Ingesting certain strains of medical marijuana with high in CBD's and low THC's and taken under the tongue in oil form before and after a concussion can decrease the brain injury recovery time and possibly reduce the risk of dementia later in life. These studies and the development of other new brain protective methods have been brought to the forefront as concussions are becoming more prevalent as NFL athletes are getting bigger, faster, and stronger . . . and hitting harder.

NFL Commissioner Roger Goodell doesn't rule out the use of medical marijuana by players for pain, prevention and treatment of head injuries. Fifty states as of 2019 passed medical marijuana laws and some with CBD only legislation which does not give patients the THC high that's usually associated with marijuana use. Medical marijuana is still illegal under the NFL's collective bargaining agreement, although the NFL has relaxed the THC limit for a players who test positive. NFL commissioner Roger Goodell left the door open for medical marijuana use when he was recently asked about the subject before the 2014 Super Bowl. Specifically, Goodell was asked if there would ever be a time when players would be permitted to use medical marijuana in states where it's legal. Instead of saying no, Goodell offered the following: "I don't know what's going to develop as far as the next opportunity for medicine to evolve and to help either deal with pain or help deal with injuries, but we will continue to support the evolution of medicine,"

Seattle Seahawks coach Pete Carroll agrees with the notion that the NFL should look into medicinal marijuana as a means of taking the best possible care of its players. While not explicitly coming out in favor of it, Carroll answered a question about NFL commissioner Roger Goodell's recent comments on the topic by making it clear he thinks it's an avenue worth pursuing. Caroll's comments were also made prior to Super Bowl XVIII in Giants Stadium.

Pete Carroll thinks that the NFL should consider medicinal marijuana as a way to treat injured players. He went on to say, "regardless of what other stigmas might be involved, we have to do this because the world of medicine is doing this." But marijuana remains on the NFL's banned substances list.

https://instagram.com/p/vWpZWNhwCp/?taken-by=petecarroll

Removing marijuana from the league's banned substance list likely would be difficult because the federal government and interstate commerce laws still consider it illegal and prohibit it from being transported across state lines. Moving forward, the NFL, Under Armor and General Electric have launched The Head Health Initiative. It is a four-year, $60 million study to speed the diagnosis and improve treatment for mild and traumatic brain injuries. The goal of the program, guided by healthcare experts, is to improve the safety of athletes, members of the military and society overall. Head Health Challenge I: Methods for Diagnosis and Prognosis of Mild Traumatic Brain Injuries was launched in March 2013 with an investment of $20 million. Winners were announced in Jan 2014. Head Health Challenge II: Innovative Approaches for Identifying and Preventing Brain Injury was launched in Sept 2013 to spur new innovations and materials that can help protect the brain from traumatic injury. Winners of Head Health Challenge II were announced Nov. 13, 2014. Head Health Challenge III was launched Feb. 14, 2014 with the NFL, GE and Under Armor joining forces with the U.S. Department of Commerce's, National Institute of Standards and Technology (NIST). Applicants submitted abstracts of products they are developing to reduce brain injuries. Winning applicants were contacted and asked to submit a full proposal and a sample of their cushioning material by June 17, 2015. Announcement of finalists was scheduled for September, 2015. The six finalists received up to $250,000 in funding.

Some NFL players are smoking marijuana, some apparently with low THC doses before and after the games. NFL players feel marijuana helps them deal with pain and possibly concussions, also verified by studies mentioned on page 33. There is a risk that players using marijuana for this purpose may fail the NFL marijuana test. The NFL has recently increased the threshold limit for a positive test on marijuana from 15 ng/ml (nanograms per milliliter)to 35 ng/ml which will help players deal with suspensions. To be really safe from suspensions there are CBD (PhytocCannabinoid) products made from industrial hemp with very little THC that can help deal with pain and concussions, with no fear of failing the NFL marijuana test and you don't have to smoke it. NFL players do not need a medical marijuana card to purchase these products and in oil form. The CBD oils made from industrial hemp can be ordered online. Go to our website for further details:

marijuanaingardenofeden.com

https://images.search.yahoo.com/images/view

Sources: www.headhealthchallenge.com

http://www.cbssports.com/nfl/eye-on-football/24402661/roger-goodell-doesnt-rule-out-medical-marijuana-use-by-players

http://espn.go.com/nfl/playoffs/2013/story/_/id/10360723/pete-carroll-seattle-seahawks-coach-says-nfl-consider-medicinal-marijuana-treat-injured-players?src=mobile#13035|304|300|ca-espn-q-story|nfl

http://www.nfl.com/news story/0ap1000000238256/article/nfl-under-armour-ge-launch-head-health-challenge-ii

Magu: The Hemp Goddess

Photo by Zbigniew Dolinski, National Museum, Warsaw, with permission

In various cultures in Asia hemp has long been associated with medicinal and spiritual practices. A prominent religion in ancient China that valued hemp was Taoism. The Chinese had a caretaker for hemp, her name is Magu. Magu was considered to be equivalent of the divine, as hemp has long been named an "elixir of life".

Goddess Magu's association with hemp primarily lies in its use as a healing plant. The majority of Magu's mythological stories revolve around the ways in which she aided the poor and the sick either as a goddess or as a priestess of a healing deity. Magu takes on a more divine role in the ancient literature in Korea.

The **National Museum in Warsaw** is one of the largest museums in Poland. It comprises a rich collection of ancient art including Egyptian, Greek, Roman, Italian, French, Italian, Dutch, German and Russian. In 2015 they had an exhibit of Oriental art, which was the largest collection of Chinese Art in Poland. They had several scrolls of Magu during their exhibition.

This particular MaGu painting, painter unknown, uses watercolors on silk. Courtesy Mikolaj Machowski, National Museum Warsaw.

Cannabis on the Brain?

Study finds cannabis combats brain degeneration and increases stamina.

The University of Bonn, Germany, has conducted a new study at their institute of Molecular Psychiatry which found that cannabis triggers the release of anti-oxidants, acting as a cleansing mechanism. This results in the removal of damaged cells and improves the efficiency of mitochondria, the energy source that powers cells, potentially increasing stamina. The findings indicate the benefits of cannabinoids can be huge, especially for those suffering from degenerative conditions effecting the brain. The discoveries highlight new data on how natural marijuana cannabinoids hold the capacity to stop brain inflammation responsible for causing cognitive decline, neural injuries or failure, and brain degeneration. Patients may be able to overcome brain conditions like Alzheimer's disease, Parkinson's disease, Huntington's disease, and more by supplying these receptor sites (endocannabinoids) with phyto cannabinoids found in marijuana and hemp plants.

Encouraged by this discovery and the possible use of cannabinoids, Dr. Gery Wenk, a professor of neuroscience, immunology and medical genetics at Ohio State University, indicated he had been trying to find a drug that would reduce brain inflammation and restore cognitive function on rats for over 25 years. He stated "cannabinoids were the first and only class of drugs that have ever been effective".

Source: Cannabis on the Brain?
www.CannaKitchenResearch.com
July/August 2013-Page 11

Who Should Not Use Marijuana?

There are unwanted risks and side effects with marijuana. Extensive studies have shown risks. Adolescents should not use marijuana or people with a genetic predisposition toward psychosis. Pregnant women and people with heart disease or weakened immune system should also not use marijuana. Heavy long term use of cannabis may lead to addiction; and long term use of smoking cannabis may lead to lung disease. Source Medical Cannabis Primer p. 113

wwwmedicalcannabisprimer.com.

Low THC Can Help Brain Damage?

Extremely low doses of marijuana's psychoactive component can protect the brain before and after injury. Though marijuana is a well known recreational drug, extensive scientific research has been conducted on the therapeutic properties of marijuana during the last decade. Medical cannabis is often used by sufferers of chronic ailments including cancer and post-traumatic stress disorder, to combat pain, insomnia, lack of appetite and other symptoms.

Now Prof. Yosef Sarne of Tel Aviv University's Adelson Center for the Biology of Addictive Diseases at the Sackler Faculty of Medicine says that the drug has neuroprotective qualities as well. Prof. Sarne has found that extremely low doses of THC—the psychoactive component of marijuana—protects the brain from long-term cognitive damage in the wake of injury from hypoxia (lack of oxygen), seizures, or toxic drugs. Brain damage can have consequences ranging from mild cognitive deficits to severe neurological damage.

Previous studies have focused on injecting high doses of THC within a very short time frame— approximately 30 minutes—before or after an injury. Prof. Sarne's current research, published in the journals Behavioral Brain Research and Experimental Brain Research, demonstrates that even extremely low doses of THC—around 1,000 to 10,000 times less than that in a conventional marijuana cigarette—administered over a wide window of 1 to 7 days before or 1 to 3 days after the injury, can jumpstart the biochemical processes which protect brain cells and preserve cognitive function over time.

This treatment, especially in light of the long time frame for administration and the low dos age, could be applicable to many cases of brain injury and be safer over time, Prof. Sarne says.

Conditioning the Brain

While performing experiments on the biology (properties) of cannabis, Prof. Sarne and his fellow researchers discovered that low doses of the drug had a big impact on cell signaling, preventing cell death and promoting growth factors. This finding led to a series of experiments designed to test the neuroprotective properties of THC in response to various brain injuries.

In the lab, researchers injected mice with a single low dose of THC either before or after exposing them to brain trauma; a control group of mice sustained brain injury but did not receive the THC treatment. When the mice were examined 3 to 7 weeks after the initial in jury, recipients of the THC treatment performed better in behavioral tests measuring learning and memory. Additionally, biochemical studies showed heightened amounts of neuroprotective chemicals in the treatment group compared to the control group.

The researchers conclude that the use of THC can help to prevent long-term cognitive dam age that results from brain injury. One explanation for this effect is pre- and post-conditioning, whereby the drug causes minute damage to the brain to build resistance and trigger protective measures in the face of much more severe injury, explains Prof. Sarne. The low dosage of THC is crucial to initiating this process without causing too much initial damage.

Preventative and Long-term Use

According to Prof. Sarne, there are several practical benefits to this treatment plan. Due to the long therapeutic time window, this treatment can be used not only to treat injury after the fact, but also to prevent injury that might occur in the future. For example, cardiopulmonary heart-lung machines used in open heart surgery carry the risk of interrupting the blood supply to the brain; this drug can be delivered before hand as a preventive measure. In addition, the low dosage makes it safe for regular use in patients at constant risk of brain injury, such as epileptics or people at a high risk of heart attack.

http://www.aftau.org/ site/News2?id=18625

Greening Out

Drinking alcohol and smoking marijuana is commonly referred to as greening out. THC, the main ingredient in marijuana, is absorbed into the blood faster when alcohol is also present. So don't drink heavily and then smoke pot. The magnified effects of using these substances together can be very unpredictable and may cause panic, anxiety, paranoia, flawed or fatal decisions. Also when recreational drinking and smoking marijuana are combined, it is easier to drink excessively and risk alcohol poisoning, which can possibly cause death. Usually, when a person drinks too much alcohol, the body responds by vomiting in order to rid itself of the excess alcohol. Since marijuana helps prevent vomiting, the body may be unable to rid itself of dangerous toxins and the a individual may choke on the vomit that is produced. Little research has been done on combining alcohol and marijuana, but the indications are that there can be major risks. No matter what drugs are used always use them responsibly.

https://www.psychologytoday.com/us/blog/ending-addiction-good/201405/the-dangers-combining-alcohol-and-marijuana#

The History of Hemp in Norway

by **Jan Bojer Vindheim**

The first certain proof of hemp in Norway is from the Viking age. Woven hemp fibers were an important pre-condition for the long sea journeys of the Vikings. Hemp was an important item of trade as bounty for their regular armed raids. The best hemp material find from Viking times were seeds found in a leather pouch of the priestess of the Norse Goddess Freya -usually called the Oseberg Queen. In the ninth century there were active trade routes both eastwards through Russia and westwards along the European coasts and waterways. Around the year 1000 we may assume that hemp was grown at several places in Norway. At the end of these medieval times hemp textiles were common. From the 13th century the King taxed hemp and from the 16th Century there are records of state income from farmers growing hemp.

In Norway, hemp fiber was mainly used as the raw material for making ropes and rough textiles. Until the end of the last century boat and ship lines and nets were made from hemp. Around the year 1000 imported Italian hemp varieties provided a better quality fiber and comparable to linen textiles. Until around 1880 it was usual for Norwegian households to spin hemp-yarn. There are reports of hemp cultivation as far north as Nordland. There are also a few reports that hemp material was used as a popular medicine.

Traditionally the male plants were harvested first and the female plants a little later. The plants were pulled up with the roots and dried in bundles. The seeds could then be shaken off before the plant stalks before soaking in water for retting. When the soft parts of the plant have been dissolved, the stalks are dried once more and pulled bit by bit through the hampebraak. This has a "mouth" of tight rollers that breaks the stalks, and makes it possible to remove the fibers or bast from the inner wooden parts or hurd. This is heavy work that was usually done by the menfolk. The next step was pulling the fibers through the hackle, a kind of comb that separates the two different fibers. This was considered lighter work and was done by the women. The hackled long thin fibers could be spun and then woven into clothing, and larger long fibers were used for making cordage or rope.

On the Danish island of Fyn, tradition holds that the art of growing hemp was brought back by sailors from the Baltic lands. (Brøndegaard 1979) In Denmark the conditions for growing hemp were much better than in Norway. King Christian IV, in 1629, demanded the Danish farmers to grow hemp to supply his navy with rope.

In Norway a law on hemp cultivation was also passed to stimulate hemp-production as 500 tons of hemp was also imported to Bergen during the period 1650-1654, mainly from the Baltic countries.

At the beginning of May, the farmer sows hemp seeds when the soil is dry and harrowed down with a wooden harrow and then rolled. Once the hemp is sown in 3 months the male plants must be pulled, or

separated from the female plants. The female hemp plant, the one that bears the seed, should yet remain standing 3 to 4 weeks to ripen before hand harvesting. (Adresseavisen 1802, nr 14)

The Danish-Norwegian kings took steps to organize and tax the rope making industry from the seventeenth century. To make the hemp ropes strong and flexible they had to be impregnated with tar. The tar was heated in large vats through which the rope could be pulled or dipped. A rope walk was therefore a fire hazard, and was for this reason often was placed outside the town center.

One single sailing ship needed several miles of various types of cordage. The pinnacle of Norwegian naval rope making was the anchor cable for the frigate Kong Sverre, made from 4 tons of Russian hemp. It was carried through the streets of Tønsberg, in the year 1864, by 120 sailors led by a marching band.

After World War I, the demand for hemp cordage dwindled swiftly. The amount imported in the period 1920-24 was only half of the amount a decade earlier. However in the late 1930s an increase was caused by the upsurge in Antarctic whaling. The whalers were fussy about their equipment, and the harpooners were especially careful about the "forerunner", a slender rope connected directly to the harpoon. Only hand spun Italian hemp had the desired quality.

With the demise of large sailing ships production shifted to other fibers and by 1960 there was little cultivation of hemp in Norway. In 1964 with new drug legislation, hemp cultivation became illegal. In the 1990s, hemp cultivation was reintroduced in several northern European countries, but a side-effect of the

hard line drug policies in Norway and Sweden made growing hemp Illegal.

In the summer of 2001, however, a public controversy arose when a small number of farmers in the county of Vestfold turned over 3000 illegally cultivated low-THC hemp plants to the police, and demanded to be prosecuted. The Minister of Agriculture immediately promised to look into the restrictions on industrial hemp, and a few weeks later his Ministry delivered a report on the problems of separating intoxicating hemp from industrial hemp strains. The ministry did not foresee any large problems in this regard, and recommended a system of licensed growing of industrial hemp.

This article was previously published in The Journal of Industrial Hemp published by the *International Hemp Association. It has been condensed for publication here and with permission of Jan Bojer Vindheim. See his complete story at www.vindheim.net/hemp.html*

Cannabis found near old Viking Settlement in Newfoundland

Vikings settled in northern Newfoundland, the site of L'Anse aux Meadows, around A.D. 1000. Recently cannabis pollen was discovered near the settlement. Researchers wondered whether the Vikings, while exploring in North America, were smoking, eating or using it to make clothing. Probably all three.

www.livescience.com - search: Archaeology
https://news.yahoo.com/were-vikings-smoking-pot?

Brief History of Hemp

Hemp is one of the earliest known cultivated plants here on planet earth. It could be inferred that hemp evolved, or was created, with mankind's earliest presence; and that could apply to recent discoveries of 900,000 year old human footprints found in the UK. Recent history and archaeological evidence attests to hemp presence around 10,000 years ago in China and the Middle East. In fact, one of the oldest known human artifact is a piece of 8000-year-old hemp fabric found in Mesopotamia and mentioned in Assyrian scripts. Please note that more on the history of hemp is covered in the Marijuana history section. In other words, in the beginning, hemp and marijuana were possibly one and the same.

Industrial hemp has been grown and cultivated in Canada since 1606, as it was one of the first crops brought to Nova Scotia by Samuel de Champlain's botanist, Louis Hebert. The Spaniards also brought hemp to South America and the new world around 1845.

The English settlers brought hemp to Jamestown in 1611 where it spread to Kentucky in 1775, then on to Missouri and Illinois. Hemp production rapidly flourished in these states between 1840 and 1860.

Harvesting Hemp for fiber in Kentucky, early 1900's -Photo from James Lane Allen's The Reign of Law, (New York, MacMillan & Co., 1900)

Over time, increased production of cotton and cheaper imports of jute competed with hemp cultivation causing a decline in hemp production.

Oregon Hemp farmers in 1891 bring their hemp to Portland, Oregon's F.S. Harmon & Co. Furniture for making fabric for their furniture.

In the early 1930's the word marijuana emerged in Mexico with reference to high THC cannabis plants. Both marijuana and hemp have separate male and female plants with pollen and flowers. Both marijuana and hemp plants have been bred into hundreds of varieties depending on soil and climate variations, geographic location and preference of market demands.

The flowers of the female hemp plant are allowed to mature into seeds for food and oil. The taller hemp stems produce long outer fibrous material or bast and a pithy spongy center material identified as the hurd. Sometimes hemp plants are harvested at different times for seed and for fiber. Depending on the variety of the plant, where it is grown and its primary intended use. The hemp seeds and stalks may be harvested together or separately.

Hemp Harvesting in UK Courtesy of American

Hemp Drying in UK Courtesy of American Lime

The very nutritious hemp seeds can be eaten milled or pressed into oil for human consumption. Hemp seeds are imported into the U.S. mostly from Canada and Europe after they are milled into smaller pieces, often referred as nutritious 'hemp hearts'. Companies--mostly in Canada--will hot or cold press the hemp seeds into oil and processed seeds for export into the United

Hemp Baling in UK Courtesy of American Lime

States. Some hemp for seed and fiber is now growing in the United States. Hemp oil and milled seeds in the U.S. can then be processed into cosmetics, hemp milk, hemp ice cream, hemp yogurt, hemp flour, hemp protein powder, hemp beer, hemp burgers and many other products.

Hurd and fiber can be used to produce paper, textiles and clothing, rugs, fuel pellets and building materials such as Hemcrete (lime and hurd), animal bedding, oil absorbent pellets, mats and insulation.

Animal lovers prefer bedding made from hemp hurd due to it's high absorption characteristics. It is a common practice to mix hemp bast fibers with cotton fibers to make quality clothing. The bast fibers can be made into strong organic plastics using a resin for body shells, door panels and other automobile components.

At present of the worldwide, hemp seeds command the major portion hemp market. Most hemp fiber production takes place in China, Europe and Canada.

Farmers can grow the desired hemp variety for the intended market. Europe, Asia, Canada and South America were early producers of hemp. Germany, the UK, Austria, and Switzerland again started producing hemp in 1990. American Lime Technology started supplying hemcrete for the construction of commercial and residential buildings in the year 2008.

Hempcrete is a lime-based blend of lime and other minerals, binder and hurd mix.

Spraying of Hempcrete for building construction. Courtesy of American Lime

Tradical Hempcrete® Courtesy of American Lime

Canada again started to grow hemp on March 12, 1998. The Canadian production and cultivation of industrial hemp has been legalized under license from Health Canada, under Bill C-81. There are many varieties of hemp grown in Canada, such as CRS1. Interestingly, many hemp varieties came from the Ukraine. In central Canada, hemp seeds are usually harvested first by a mobile harvester (combine) that clips the seed tops from the hemp plant leaving the stalks with long fibers or bast and spongy or hurd fibers for retting (allowing the hemp stalks or stems to break down with added moisture and weathering) to be harvested later. After a month or so of allowing the retting to occur, hemp farmers gather the hemp stalks, usually with a round or square baler, and move their crop to a storage shed for future processing. In more humid climates or with certain varieties, the whole hemp plant is harvested at one time for seed, leaving the stalks in the field after harvest for retting and baling.

BMW Door panels (from hemp fiber) Courtesy of American Lime

Chinese farmer harvesting hemp. https://www.facebook.com/HempTraders/photos/a.108270112573065.7442.108103129256430/108270289239714/?type=3&theater

China, is the world's leading hemp producer and exporter. Other countries producing hemp include: Canada, France, Chile, Russia, North Korea, Turkey, Ukraine, Uruguay, Romania, Hungary, Poland, Austria, Spain, Serbia, UK, Finland, Australia, United States and Canada. Hemp production in these countries is increasing the GNP though jobs and exports.

With the passage of the 2018 Farm Bill, hemp was removed from the DEA's Controlled substance list which included Heroin, LSD, Ecstasy and others. Farmers can now legally and grow hemp and businesses can legally process hemp into a variety of health foods, medical and fiber products. The projected value of hemp foods, oil, health supplements, pharmaceuticals, fiber for textiles, clothing, absorbent materials, building and construction materials imported into the U.S. primarily from Canada and China to be around 2 billion dollars annually. The **United States is catching up in hemp production**, primarily CBD hemp, and soon will reach 2 billion in sales annually.

Private Housing built with Tradical Hemcrete®
Courtesy of American Lime

Workers at China hemp factory fold textiles hemp-cotton fabrics for shipment to garment factories and distributors worldwide.
Courtesy of http://www.c-hemp.com/about.html

Art Potoroka, hemp farmer near Ethelbert, Manitoba shows his hemp plants in late summer.

Breathe® Insulation (from hemp fibre)
Courtesy of American Lime

Art Potoroka harvesting hemp near Gilbert Plains, Manitoba, Canada with a seed producing variety, CRS1

2018 FARM BILL

The 2018 Farm Bill was signed into law on December 20th, 2018 and went into effect on January 1st, 2019.The 2018 Farm Bill amends the federal Controlled Substances Act (CSA) of 1970 so that hemp plants containing 0.3 percent THC or less are no longer classified as a schedule I controlled substance under federal law.

The CSA acknowledges two types of cannabis, hemp and marijuana. Hemp is an agricultural commodity. Marijuana is a controlled substance. The 2018 Farm Bill removes hemp from the jurisdiction of the Justice Department, specifically the Drug Enforcement DEA. moves hemp to the U.S. Department of Agriculture.

- The 2018 Farm Bill permits those US states that wish to grow hemp to submit a plan to the US Secretary of Agriculture. But it allows all tribal lands to grow hemp with or without a state sanctioned program.
- The 2018 Farm Bill extends federal crop insurance coverage to industrial hemp. Hemp producers can apply for USDA certification and grants, like other agricultural commodities. States that don't adopt an industrial hemp program cannot interfere with the transportation or shipment of industrial hemp through their states.
- The FDA immediately issued a statement saying, "We treat products containing cannabis or cannabis-derived compounds as we do any other FDA-regulated products -- meaning they're subject to the same authorities and requirements as FDA-regulated products containing any other substance." The FDA went on to state, **"It's unlawful under the FD&C Act to introduce food containing added CBD or THC into interstate commerce, or to market CBD or THC products as, or in, dietary supplements, regardless of whether the substances are hemp-derived. This is because both CBD and THC are active ingredients in FDA-approved drugs and are subject to substantial clinical investigations before they are marketed as foods or dietary supplements."**

"This law marks the first change in the federal classification of the cannabis plant since it was initially classified as a Schedule I controlled substance by Congress in 1970, and paves the way for the first federally-sanctioned commercial hemp grows since World War II." – Paul Armentano, NORML.

Farm Bill 2018 – Analysis of Hemp-Specific Provisions – Foundation for an Informed Texas

CBD industrial hemp field in the Willamette Valley of Oregon

US STATES INVOLVED WITH INDUSTRIAL HEMP

As of the present time 44 states have passed legislation to now allow industrial hemp farming for research and/or commercial purposes, allow university research; or have laws pending . . Alabama, Alaska, Arizona, Arkansas, California, Colorado, Connecticut, Delaware, Florida, Hawaii, Illinois, Indiana, Kansas, Kentucky, Maine, Maryland, Massachusetts, Michigan, Minnesota, Missouri, Montana, Nebraska, Nevada, New Hampshire, New Jersey, New Mexico, New York, North Carolina, North Dakota, Oklahoma, Oregon, Pennsylvania, Rhode Island, South Carolina, South Dakota, Tennessee, Texas, Utah, Vermont, Virginia, Washington, West Virginia, Wisconsin, and Wyoming.

HEMPCRETE ECO LEGO BUILDING BLOCKS

Industrial hemp is a tall plant that when harvested for seeds the stalks can be decorticated by a machine that separates the outer long fiber called bast from the inner short spongy fiber of the hemp plant called hurd. When this hurd is mixed with hydraulic lime and water and dries it creates great insulating and breathable hempcrete structural walls or insulation for houses. The walls or insulation also breathe in CO2 from the atmosphere, depositing small pieces of carbon over time that makes the wall stronger and can last for centuries.

The hempcrete walls do not need a vapor barrier as it breathes in just the right amount of moisture to make the living area more comfortable and prevent moisture buildup in the walls. Vapor permeable is the word. In other words: hempcrete walls regulate indoor humidity and temperature. The lime binder provides fire and mold resistance, repels pests, and makes hempcrete extremely durable.

A renaissance in hempcrete homes are emerging worldwide. They are sprouting up in UK, France, Australia, Canada and the US. In the UK, Canada, Europe and U.S. many homes are being built with hempcrete. Just BioFiber Structural Solutions of Alberta Canada has recently completed Arno Keinonen's house on Vancouver Island with interlocking high load bearing hempcrete building blocks. These blocks can be used as a stand alone load supporting wall. Mac Radford from Calgary, Alberta invented the hempcrete building block having been inspired by Legos. It took Mac several years to develop these load bearing interlocking building blocks.

At a resent Hemp Building Symposium in Quebec Canada, Terry Radford, President of Just BioFiber Structural Solutions, and brother, Mac, show-cased their revolutionary building product. Mac named their hempcrete blocks, 'eco lego blocks'. These eco lego blocks with interlocking pegs and ends can make construction more attractive to city planners and building code officials as their eco lego blocks have been standardized. The problem with hempcrete construction, right now, is each hempcrete home has to be inspected to get an exemption from the building code, thus It's difficult for hempcrete builders to get their building permits approved for a mortgage.

Just BioFiber Structural Solutions and Ridgeco Developments, the builder, recently completed a 4,000-square-foot home near Sooke, British Columbia. Jack Anderson of Greenspace was the architect. Other eco lego block homes are under construction in western Canada.

A HARMLESS HOME

by Arno Keinonen, Owner

It has taken quite a while to design and build a sustainable pioneering home for working and living a simple life. whilst also being able to grow food within the home. Our main target for this pioneering home was to follow the principle: **First do no harm** After initial contemplation of how to do this we realized that the process of building a home, consuming various materials and living in a home will cause many kinds of unavoidable harm. The best we can achieve is to identify the types of consumption and harm, establish a way to assess/ measure these and to do our best to minimize the disturbance we cause to nature and our healthy living environment. We call our project: **A Harmless Home** We decided to focus on at least the following: 1. reducing consumption of natural resources – for construction and during the lifetime of the home. 2. making the home a healthy and comfortable place for the inhabitants. 3. making building the home economical and feasible to adopt by the home construction industry. The Harmless Home is complete, built on a rocky knoll with a great view of the Jaun de Fuca Strait and Olympic Mountains. The home is located a few miles west of Victoria, BC near Sooke, BC. The project is intended to be fully public. http://www.harmlesshome.ca

Arno Keinonen, owner, and Mac Radford, inventor of hempcrete eco lego blocks.

Harmless Home under construction.

Southside view, with interior greenhouse in center with skylights; and solar panels on roof.

Northside view of the Harmless Home

BMW i3 interior hemp related upholstery

BMW i3 interior upholstery and door panels made from natural Kenaf and hemp fibers.
Photo by Bruce Meland

The interior of the BMW i3 also features use of natural KENAF fibers. Its characteristic fiber structure, underscores the premium claim of the BMW i3.

http://www.bmwusa.com/Standard/Content/Vehicles/2014/BMWi/default.aspx

CANADIAN COMPANY MAKES HEMP FIBER FOR AUTOMOBILES

In 1941 Henry Ford constructed a prototype car made from hemp fiber. He mixed the long hemp fiber with resins and heated the mixture in a formed press, depending what car body part he was making. Henry Ford sledge hammered the back trunk and could not make a dent. It has been known that plastic and carbon fiber body parts are many times lighter than steel, but auto manufactures are now realizing that hemp fiber composite auto parts are even lighter and much stronger than present plastic or carbon fiber body parts.

Fast forward to today: A Canadian company, Bio Composites Group Inc., in Drayton Valley, located in north central Alberta, is making hemp fiber mats which are sent to automotive companies who mix resin with the hemp fiber and heat in a mold to make extremely strong and light weight door panels and other body parts. Bio Composites Group makes erosion control matts and Terafiber grow mats from hemp fiber which are ideal for sprouts, wheat grass and other short rotation crops.

www.biocompositesgroup.com

Henry Ford sledgehammers his 1941 Coupe with hempcrete fiber composite body; and no dent.

https://postmediadriving.files.wordpress.com

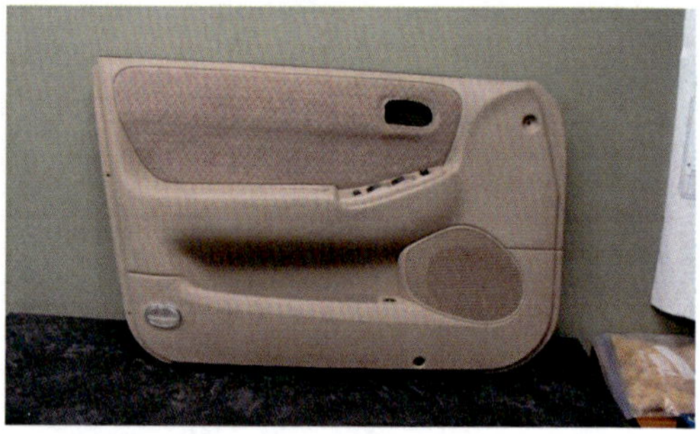

BioComposite Group ships hemp fiber to American auto companies for prototype door panel testing

Hemp Clothing Is Good For Your Health

Staph infections worldwide, mostly at medical facilities, are a major killer. Hemp clothing is a number one defense against staph infections. China's military is continuing to use more and more hemp clothing to fight this dreaded disease. It is estimated that each year, 2 million Americans become infected during hospital stays - - and at least 90,000 of them die.

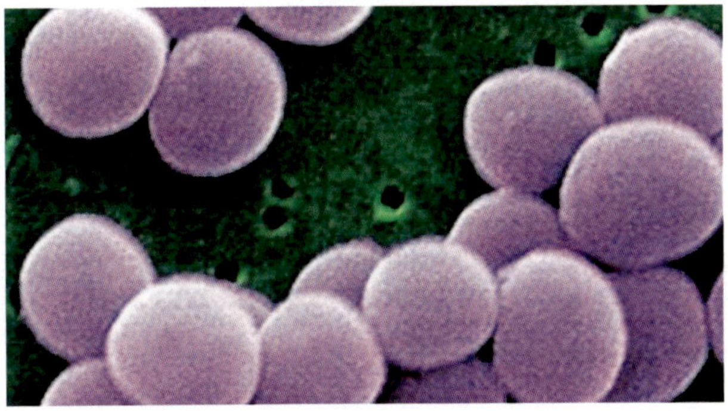

Staphylococcus aureus www.evirotextile.com

MRSA (an antibiotic resistant strain of staph) is a leading cause of hospital-borne infections. Staph is spread by direct contact: touching items that are contaminated such as towels, sheets, privacy curtains, and clothing One of the most important recent discoveries in the textile industry is the ability of hemp fabrics to kill surface bacteria while synthetic or man made polyester, and polyethylene fabrics allow it to remain on their surfaces for up to several months at a time.

Unknown to many, hemp fabrics exist in today's market that can replace each of these transmission prone hospital fabrics. Technological improvements, to say the least.

In hemp textile development began in the early 90s when EnviroTextiles lead textile engineer, Barbara Filippone, began working with hemp fabrics in China. To date, the company has over 100 hemp and hemp blended fabrics available to suit any traditional fabric application. In addition to staph resistance, 98.5% of the fabrics were bacteria free during the first measurement of the testing. Other tests show that hemp fabrics have superior resistance to UV and infrared wavelengths, providing multiple applications for military and hospital use worldwide.

Many of these hemp fabrics can also help the general population, preventing the spread of diseases world wide. EnviroTextiles' hemp fabrics have recently been approved by the USDA as Federally Preferred for Procurement under their BioPreferred Program. EnviroTextiles hemp fabrics, when blended with other natural fibers, are also flame resistant - unlike most man made or synthetic fabrics. Envirotextiles, LLC is a woman-owned industrial hemp and natural fiber manufacturing company with headquarters in Glenwood Springs, CO.

Ladies Hemp Hoodie www.envirotextile.com

It is the largest importer of hemp and natural fiber textiles and related products in the United States. EnviroTextiles proudly sells their products in the U.S. and in over 70 countries worldwide. The company presently has their presence in the US, China, and Mexico, and focuses on natural fiber resources and economic development in regions at the community level.

References:

1. www.envirotextile.com
2. Survival of Enterococci and Staphylococci on Hospital Fabrics and Plastic –http://www.ncbi. nlm.nih.gov/pmc/articles/PMC86187/

HEMP DECORTICATOR IN NORTHERN ALBERTA

A decorticator plant is a machine or a series of machines that separate fiber from hemp, flax, and canaf into two different fibers.

1. The bast fibers are the long very thin fibers from the outside of the stalk or stem, that when decorticated or separated, the fibers can be used to make clothing, very strong bio composites (used in making door panels and other body parts for automobiles), papers, and geotextile matting to name a few.
2. The hurd, the inner short pithy absorbent fibers of the stalk or stem can be made into hempcrete (a concrete replacement), building materials similar to OSB for walls, insulation, or highly absorbent animal bedding to name a few.

Hemp is one of the fibers processed at InnoTech Alberta in Vegreville, Alberta, Canada. They have set up a pilot scale decortication facility to primarily process hemp fiber for the biocomposite and construction industries. Most of the hardware comes from Belgium. Agricultural producers, private businesses, and governments have utilized this decortication facility to demonstrate success in creating fiber products from hemp or flax before building a commercial decortication plant. For example, the facility has been utilized to process hemp for use in a building material called hempcrete. Essentially it is a research and development facility and the decortication equipment can be adjusted to meet the client's needs in developing their very special product such as Just BioFiber's hempcrete building blocks which uses hemp hurd. The Biocomposite Group uses the bast fiber for making hemp biocomposite body parts for automobiles, absorbing mats for oil spills or erosion control. It gives clients flexibility in terms of what stage in the complete production process it is used. Material can be added or removed at different points, depending on the client's needs.

A 1,000-pound bale of hemp produces approximately 200-250 pounds of bast fiber (depending on straw quality), 500 pounds of Hurd fiber, 150 pounds of dust, and 100 pounds of mixed short-fiber and hurd.

Decortication Plant in Vegreville, Alberta, Canada

Contributed by Byron James www.albertainnovates.ca

SMALL PORTABLE DECORTICATOR

Hill Agra Machinery of Melancthon, ON, Canada has been building portable decorticators. They can go into the hemp field after harvest and decorticate the hemp stalks into long fiber, Bast and short pithy fiber, Hurd. Advantages of field decortication; (1) don't have to haul bulky hemp bales to storage, (2) dust is discharged to the open air and not in a building where dust has to be discharged or bagged. The unit is hydraulically driven so a tractor can take it to the field as well as power the machine. It has 15 twin counter rotating rollers to separate the long fiber or Bast from the short spongy fiber or hurd. The hurd falls out the bottom into a side discharge conveyor and fiber out the back onto discharge chute. www.hillagramachinery.com

Hill Agra Decorticator

EARTH FRIENDLY HEMP POLYMERS

The Hemp Plastic Company is on a mission to revolutionize the plastics industry and change the landscape for the future by providing alternatives to petroleum-based plastics. With their eco friendly HempPropylene, they now have the products with renewable materials that are far easier on the earth, and at a comparable price to their petroleum-based counterparts. Now companies can switch to a greener, more eco-friendly option without compromising the bottom line.

Their biopolymer products are renewable, sustainable and, in some cases, compostable. Their unique process separates and matches different features of the hemp plant to create hemp-based polymers that can be used in virtually every facet of the plastics industry.

HempPROPYLENE

MATERIAL FOR INJECTION MOLDING

Best for: wire spools, packaging, bottle caps, automotive parts and components, gameboys, pocket combs, some musical instruments (and parts of them), one-piece chairs and small tables, storage containers, mechanical parts (including gears), toys, packaging products and most other plastic products. They can be made to meet FDA standards for food, pharmaceuticals and agricultural products.

HempETHYLENE - MATERIAL FOR FILMS AND BLOW MOLDING

Best for: parts, toys, packaging. Can be made to meet FDA standards for food, pharmaceuticals and agricultural products.

mpABS - MATERIAL FOR HIGH IMPACT, ULTRA STRONG PRODUCTS.

Best for: parts, toys, packaging. Can be made to meet FDA standards for food, pharmaceuticals and agricultural products.

HempPLA - MATERIAL FOR 3D PRINTERS

Best for: parts, toys, packaging. Can be made to meet FDA standards for food, pharmaceuticals and agricultural products.

https://hempplastic.com/manufacturers

GROWING, PROCESSING AND MARKETING CBD INDUSTRIAL HEMP IN OREGON

Present day farmers in the United States are growing mainly 2 different varieties of Industrial hemp. The tall variety of Industrial Hemp is generally grown for seed, food, and fiber; and the shorter bushy variety is generally referred to as CBD hemp. CBD hemp plants look similar to marijuana plants by size and shape but CBD hemp plants are high in various cannabinoids that have therapeutic or medicinal properties for treating diseases such as epilepsy, cancer, neurological disorders, inflammation, sleep problems, chronic pain to name a few. There can be as many as 90 different cannabinoid molecules in the CBD industrial hemp flowers, buds, leaves, and stalks when grown. Most in CBD hemp varieties are non-psychoactive and have less than 0.3% THC. (tetrahydrocannabinol)

CBD hemp plants growing in Willamette Valley of Oregon

Some states have allowed farmers to grow this CBD industrial hemp varieties for medicinal purposes. Oregon is one state that is growing CBD industrial hemp. It is more expensive to grow CBD industrial hemp than the tall hemp variety. Even though it is more expensive and time consuming, growing CBD hemp can give a greater return on their investment; as well as for processors, distributors and dealers.

At present time it takes about 1000 seeds or female starter plants generally referred to as clones, grown in a greenhouse, to grow on an acre of hemp. CBD hemp starters can be grown more intensively in greenhouses; and not have to worry about pollen from nearby marijuana or industrial hemp destroying their operation. At present the going price is 50 cents to $1 /seed and around $5 for a cloned female plant. In drier climates CBD hemp is usually grown in mounded rows with drip irrigation and covered with plastic. Most CBD hemp plants are harvested by hand where the top part of the stem of pre-flowering buds and top leaves are hand clipped and sent to a drying shed. In dry climates, field drying is possible.

Once the plants are dried they are sent to oil extracting facility. There are many different methods of extracting oil from the dried CBD plant material. Extracting processes include ethanol, methanol, alcohol, CO_2, butane and freeze extraction. CBD oil extraction is expensive and a major cost in producing CBD oil.

Small Methanol Distillation CBD Oil Extractor

The price the farmer gets for his oil is dependent on the quality of the oil. Quality is dependent on content of key CBD molecules and terpenes found in the oil. The CBD oil has to be analyzed by a licensed testing lab. Key molecules found in the CBD oil can be found on page 50 and 51. The price the farmer gets can range from $2 to $20 per gram of oil depending on market conditions and

CBD Products Go Mainstream

The passage of the Hemp Farming Act in 2018 in the U.S. allowed hemp to become regular agricultural commodity and not a Schedule 1 Drug like marijuana. Marijuana is still controlled by the Drug Enforcement Administration. Consequently, farmers have to determine what type of hemp to grow and sign up with their State Dept of Agriculture. Processors and retailers will follow suit. Word is out that CBD hemp with great potential health benefits are a more profitable commodity than the taller industrial hemp cousin. Industrial hemp seeds prices to farmers have plummeted due to oversupply worldwide. CBD hemp oil, processed from the flower of the CBD hemp plant, offers wide variety of medical and health related products and demand keeps increasing exponentially.

Hundreds of companies in the US, Canada, Israel and China are already producing CBD health-related products, infusing CBD oil into a wide variety of foods, beverages, capsules, gels, tinctures, soaps, balms, creams and topicals, to name a few. CBD tinctures have also started to show up in coffee shops and bars nationwide. These CBD hemp plants also contain terpenes, the organic compounds that give plants flavor and scent but also great health benefits. Without natural terpenes, abundantly present in the CBD flower and hence extracted for the CBD's, health benefits would be greatly reduced. With this CBD product revolution of CBD oil is infused into almost every health and beauty product imaginable, you might call it a CBD grass roots revolution, especially here in the United States.

But the FDA here in the United States says "not so fast". The 2018 farm bill removed hemp from oversite by DEA (Drug Enforcement Administration) and transferred it to the Department of Agriculture; but it did not remove oversight from FDA (Food and Drug Administration). According to recent rulings the FDA still has oversight of CBD products. FDA says CBD can't be added to food, or sold as dietary supplements or medical uses, because their officials have not determined whether CBD is safe and effective.

Republican and Democrats in Congress are urging the FDA to change it's stance. Some retailers in California, Texas, Ohio and NY have been raided for selling CBD infused products in drinks like in coffee and alcohol. Also some companies have made false claims about their products. FDA recently indicated that enforcement at present is being limited to CBD products that make false claims and/or people who have complained. Many companies are printing labels "Not Approved by the FDA" along with disclaimers about their products.
https://www.cannabisnewswire.com/cbd-going-mainstream-amid-flood-of-new-products-celebrity-endorsements-and-emerging-consensus-about-benefits/

Tetrahydrocannabinolic Acid (THC-A)

THC-A is the most common cannabinoid found in the *Cannabis Sativa/Indica* plant. THC-A is non-psychoactive and is the precursor to THC.

Tetrahydrocannabinol (THC)

THC is the primary psychoactive compound found in *Cannabis Sativa/Indica*. THC is the decarboxylated form of THC-A. THC is the most well-known cannabinoid.

Cannabidiolic Acid (CBD-A)

CBDA is one of four results of CBG-A converting into another cannabinoid. When CBD-A is decarboxylated it typically becomes CBD. It is most well-known for inhibiting cancer cell growth, reducing vomiting and nausea, slowing bacterial growth, and reducing inflammation.

Cannabidiol (CBD)

Cannabidiol is the principal non-psychoactive cannabinoid in the *Cannabis Sativa/Indica* plant. CBD has a wide range of psychological and physical effects.

Cannabinol (CBN)

THC exposed to oxygen and heat will eventually convert into CBN. CBN is mildly psychoactive but still quite potent in its therapeutic effects.

Cannabigerol (CBG)

CBG is considered the stem cell or mother cannabinoid. CBG can change into different cannabinoids and drastically change the plant and its effects. CBG is not psychoactive, acting more like an antagonist to THC. CBG is known for its anti-bacterial, anti-tumor, and neurogenesis effects.

Cannabinoid Profile

Cannabinoids help with:

Psychiatric & Neurological Disorders
Sleep Problems & Chronic Pain
Eating Disorders and GI Problems

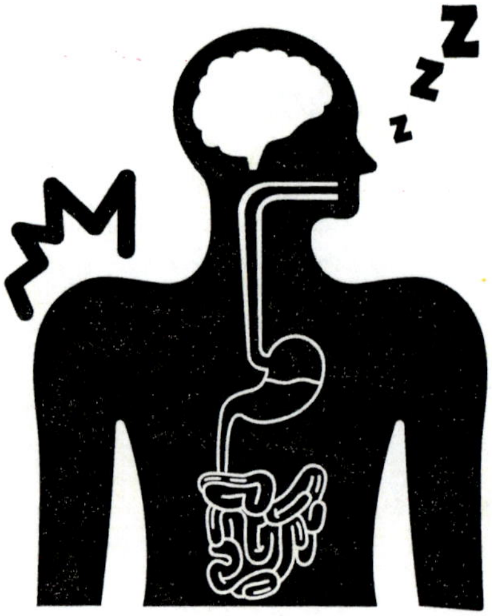

Common Terpenes found in *Cannabis Sativa/Indica*

Alpha-Pinene is found in the oils of many coniferous trees. a-Pinene is the most common terpene in the plant world and in Cannabis. a-Pinene is a bronchodilator, promotes alertness, and memory retention.

Myrcene is found in Cannabis, lemongrass, hops, and more. It is a pain reducer, anti-spasmodic, and anti-inflammatory properties. Myrcene gives off an earthy, musty quality with citrus overtones.

Limonene gives off a lemon odor and takes its name from the lemon itself. Limonene has anti-anxiety, anti-depressant, anti-biotic, and anti-oxidant properties.

Eucalyptol and camphor odor.

Linalool encompasses many herbal, minty, cinnamon, and rosewood odors. Its therapeutic effects include as an anti-anxiety, sedative, pain reduction, and anti-epileptic.

Terpineol gives off a sharp pine smell as it is related to the Lamiaceae family.

Menthol mint odor. Minor analgesic, anti-inflammatory properties.

Cintronellol is found in roses and geraniums. Citronella smell.

Carophyllene is most common in the odors of Cannabis, hops, and rosemary. Carophyllene exerts anti-inflammatory, anti-depressant, anti-alcoholic, anxiolytic, neuro-protective properties. Spicy, earthy smell.

Humulene is the most common essential oil found in hops. Humulene has excellent anti-anxiety, neurosis, and anti-inflammatory effects. Spicy green odor.

MRX Labs   FULL SPECTRUM TESTING

THERAPEUTIC EFFECTS OF CANNABINOIDS

Treating Diseases with CBD industrial hemp includes some THC molecules but less than 0.3% by weight and many more CBD molecules. These health treating CBD molecules are found in abundance in CBD industrial hemp plants. Farmers are growing CBD hemp plants which are different from their tall brother-cousin which is grown for health foods and cosmetics; derived from their seeds. CBD hemp plants appear similar in appearance and size to marijuana plants, 3-8 ft. high, but have THC content less than 0.3%. These female CBD hemp plants produce a variety of CBD molecules and a few THC molecules. When the flowering tops these CBD hemp plants are harvested and processed there can be as many as 90 different molecules extracted into oil form. Different combinations of CBD and some THC molecules exist in various varieties of CBD industrial hemp plants. The trick is to breed a CBD plant that has the best combinations for treating various diseases.

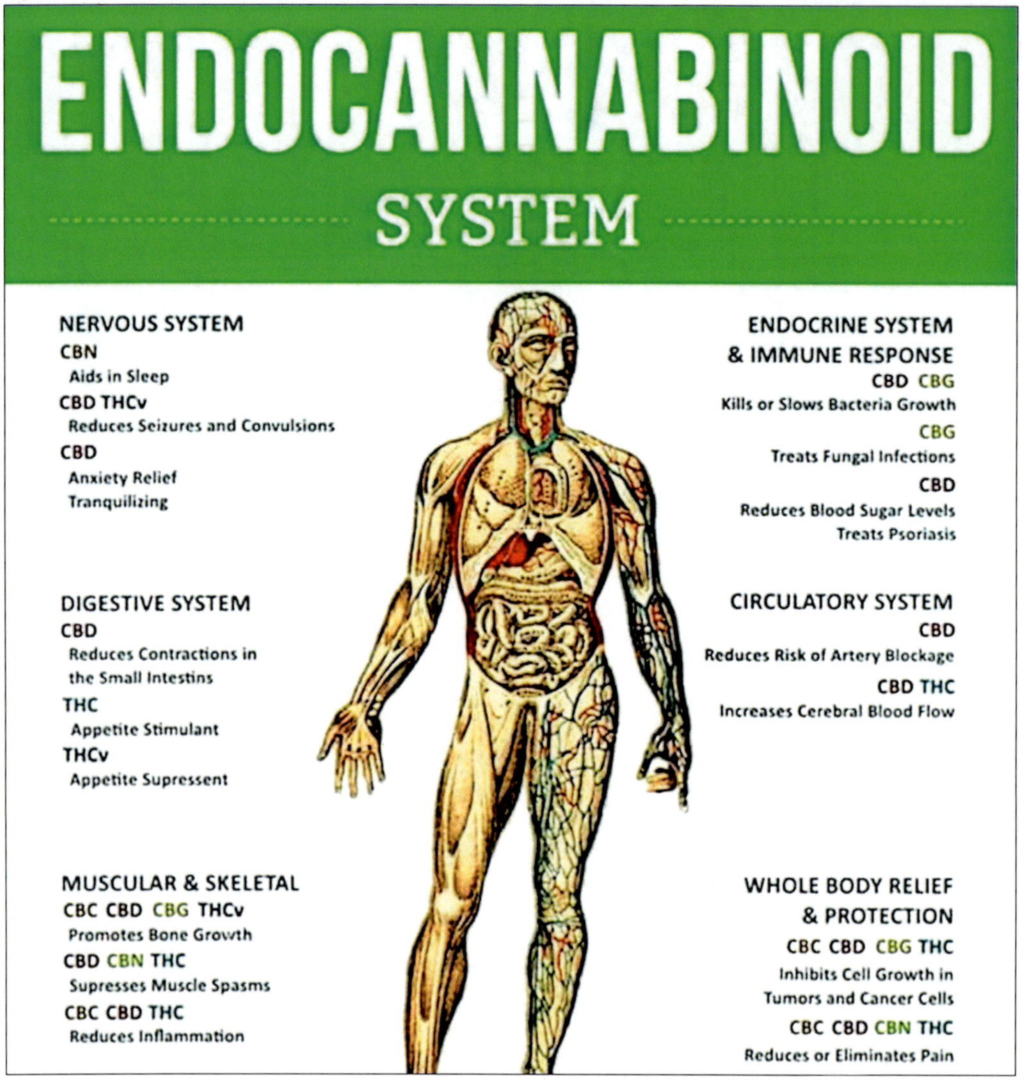

CANNABINOIDS AN OVERVIEW

Cannabinoids are the molecules that are found in the cannabis plant and created naturally by our bodies. Cannabinoids are like little keys that open pathways in cells to help them regulate a wide range of functions from metabolism to immune response. Nerve cells, like found in the brain love, THC cannabinoids. They fit into receptors in brain cells and causes a cascade of effects including upping production of dopamine — the stuff that makes you feel happy and content. CBD is still great for your brain and body but doesn't get you high.

Of all the great and powerful things that THC and CBD do to brain cells, they do to almost all the cells in your body. Cannabinoids are also very powerful antioxidants. Far more powerful than vitamin C. Antioxidants are molecules that latch onto bad molecules called free radicals that can damage cells and cause old saggy skin, and even cancer. Free radicals cause oxidation. Cannabinoids in this case, render free radicals harmless.

THC and CBD are also known to be powerful anti-inflammatory agents. Anti-inflammatories are compounds that help regulate your body's inflammation response. Inflammation goes on at a cellular level, so you can't always see it. But it's there, and cannabinoids can help minimize it. So, combine THC and CBD with other healthful and helpful ingredients such as shea and cocoa butter help the cannabinoids penetrate deeply while moisturizing your skin.
https://www.highdesertpure.com/lotion-old/

Bio of Author

Perhaps you are wondering how I got involved in the marijuana and hemp movement. I was born in Portland, Oregon, in 1939 and moved to my grandfather's farm near Silverton, OR, at age 6. We farmed turkeys and then grass seed for your lawns. I attended Silverton High School and then Oregon State University with a BS in Science and Research Meteorology in 1962. I did a Masters Theses in 1966 on grass straw disposal and marketing. Started a business operation processing, warehousing and marketing grass seed straw in mid Willamette Valley. When hemp came to Oregon in 2014, I became interested in the fiber part of hemp and joined the Oregon Industrial Hemp Farmers Association and started developing hemp fiber products. I began attending Oregon Hemp Conventions in Portland each year, and became active in networking with hemp farmers, processors and dealers.

I made a trip up to Dauphin, Manitoba, to visit hemp farmers and processors. Manitoba Harvest got started by a group of farmers around Dauphin. The founder of Manitoba Harvest started on a shoe string and local hemp farmers funded it's startup by taking stock in the company for their hemp seed delivered to Manitoba Harvest in late 90's. Manitoba Harvest was sold a few years ago and the wise hemp farmers cashed in handsomely! Many photos and stories in this book were from Dauphin, MB. That was the beginning of this book. It first came out as a e-book but not successful but it was a start.

Two years ago I decided to go soft cover with updated stories and 2 sections, Marijuana and Hemp or Marijuana and Hemp 101. The first 50 books published I took to the Oregon Hemp Convention in early April. I sold out but realized the book needed an upgrade in wording and layout and changed the title. So here it is. I received permission from authors to use their stories, interviewed businesses and individuals who had interesting projects and summarized articles from the internet. This soft cover book will be sold in some local bookstores, printed on demand by Amazon's Create Space, and from my website:

www.marijuanaingardenofeden.com

T-shirts, below, will be for sale on my website or email: etimes@teleport.coml

Bruce Meland

EPILOGUE: MARIJUANA AND HEMP

One of the reasons the FDA continues to fight the legalization of medical marijuana and CBD medical products is lack of research in the US on treatment results and side effects. Israel has gotten a jump start on cannabis research. Israel legalized medical cannabis in the early 1990's and decriminalized recreational marijuana in 2017. Israel has been doing cannabis research for over 10 years and have defined cannabis products for export to the United States. One of Israel's companies, Tikun Olam, is already selling medical cannabis products in Delaware, California and Florida.

In November 2018 Michigan voters passed Prop 1, the measure to legalize recreational marijuana. One of the advertisements during the campaign involved a trio of veterans talking about Prop 1 and how legal cannabis is important for Michigan veterans suffering with PTSD. One veteran said "marijuana is already here and it is not dangerous." Also, Michigan will get a half a billion dollars in new revenue for more funding for PTSD. According to experts in the field there has not been a new medication for PTSD approved in 17 years. Most government studies in the past for cannabis research was for drug abuse and not as a medicine. Powerful groups are trying to maintain the pharmaceutical status quo. The more cannabis looks like medicine the more it softens the public attitude about cannabis, which is very dangerous to their revenue stream. The above comments were from Dr. Sue Sisley who spearheaded a FDA approved research on whole plant cannabis as a potential treatment of PTSD for military veterans.

At present there is an oversupply of industrial hemp seeds in the United States and Canada. China is starting to export hemp seeds to the US as well. Prices have dropped significantly from 90 cents a pound to around 25 cents a pound. With a limited market for the hemp fiber, growing Industrial hemp will decline. On the other hand CBD hemp production is growing rapidly in the U.S. due to high demand for CBD oil. As a result venture capital is flowing fast to many startup CBD companies that make CBD cosmetics, foods and drinks. But again the FDA is the fly in the ointment. The FDA is trying to hold back this multi billion dollar industry. But there is light at the end of the tunnel. Congress, specifically Senate heavyweights, Sen. Mitch McConnell of Kentucky and Sen. Ron Wyden of Oregon, are to meet with FDA officials soon to hopefully solve this bottleneck.

The passage of the 2018 Farm Bill removed Hemp, which was a No. 1 Controlled Substance like heroin, from the DEA, Drug Enforcement Administration to the Department of Agriculture. But Hemp CBD products and their distribution are still under control of the FDA, Food and Drug Administration. The FDA still says more studies are needed on CBD products to make sure they are safe from side effects. Recently there have been arrests of individuals possessing CBD tinctures and topicals used for pain. The FDA is also raiding stores that sell CBD Products. Some warehouses that distribute CBD products have lost several hundred thousands dollars in confiscations. There is much confusion about these actions but until Congress forces the FDA to release their oversight on CBD products this harassment will continue.

Steve's Story - A few years ago a friend Rebecca noticed several lesions on the skin of Steve's body, a 12 year old black lab. The cancer-like lesions were gradually getting larger and his energy level was tanking. Some lesions broke through the skin leaking large amounts of puss. Rebecca had **tried some alternate treatments, including colloidal** silver, but to no avail. She had heard of a CBD oil made of CBD industrial hemp plants and sold online. She started to give Steve 6-10 drops on his tongue once a day. Then CBD drops were placed in Steve's food dish. Steve lapped it up immediately. In less than a week his energy level improved, and In a month the lesions were gone and his shiny black coat was back. For maintenance Steve would get a few drops in in his dish daily. Steve lived another active three years and passed to dog heaven peacefully.

Made in United States
North Haven, CT
05 November 2026

21765361R00031